Advances in Materials for Nuclear Energy

MATERIALS RESEARCH SOCIETY
SYMPOSIUM PROCEEDINGS VOLUME 1514

Advances in Materials for Nuclear Energy

Symposium held November 25–30, 2012, Boston, Massachusetts, U.S.A.

EDITORS

Chaitanya S. Deo
Georgia Institute of Technology
Atlanta, Georgia, U.S.A.

Gianguido Baldinozzi
CNRS-Ecole Centrale Paris
Chatenay-Malabry, France

Maria Jose Caturla
Universidad de Alicante
Alicante, Spain

Chu-Chun Fu
CEA-Saclay DEN/DMN/SRMP
Gif-sur-Yvette cedex, France

Kazuhiro Yasuda
Kyushu University
Fukuoka, Japan

Yanwen Zhang
Oak Ridge National Laboratory
Oak Ridge, Tennessee, U.S.A.

Materials Research Society
Warrendale, Pennsylvania

Shaftesbury Road, Cambridge CB2 8EA, United Kingdom

One Liberty Plaza, 20th Floor, New York, NY 10006, USA

477 Williamstown Road, Port Melbourne, VIC 3207, Australia

314–321, 3rd Floor, Plot 3, Splendor Forum, Jasola District Centre, New Delhi – 110025, India

103 Penang Road, #05–06/07, Visioncrest Commercial, Singapore 238467

Cambridge University Press is part of Cambridge University Press & Assessment, a department of the University of Cambridge.

We share the University's mission to contribute to society through the pursuit of education, learning and research at the highest international levels of excellence.

www.cambridge.org
Information on this title: www.cambridge.org/9781605114910

Materials Research Society
506 Keystone Drive, Warrendale, PA 15086
http://www.mrs.org

First published 2013

CODEN: MRSPDH

A catalogue record for this publication is available from the British Library

ISBN 978-1-605-11491-0 Hardback

CONTENTS

*Invited Paper

RADIATION EFFECTS

*Invited Paper

SYNTHESIS, CHARACTERIZATION AND THERMOMECHANICAL PROPERTIES

*Invited Paper

[illegible] CHARACTERIZATION [illegible] MECHANICAL PROPERTIES

[illegible]

PREFACE

Symposium HH, "Advances in Materials for Nuclear Energy," was held Nov. 25–30 at the 2012 MRS Fall Meeting in Boston, Massachusetts.

As the world faces steadily rising energy demand and cost, nuclear energy produced by fission and fusion reactors is increasingly recognized as an economically viable and carbon-neutral alternative to fossil energy sources. With new reactor concepts pursuing passive safety mechanisms and better performance, the behavior of structural materials and fuel is at the forefront of challenges in the development of promising reactor ideas. Accordingly, the research and development activities in nuclear materials and fuel areas have substantially increased over the last few years. The goal of this symposium was to provide a forum for the discussion of materials limitations and new developments in the field of nuclear fission and fusion energy on an experimental and modeling platform. This symposium proceedings volume represents the recent advances in materials for nuclear energy applications. The papers are divided into three sections: (1) Theory, Modeling and Simulation, (2) Radiation Effects and (3) Synthesis, Characterization and Thermomechanical Properties. Each paper in this volume provides a glimpse of the exciting recent developments occurring in materials for nuclear energy and includes the new results of both experimental and theoretical studies.

Chaitanya Deo
Gianguido Baldinozzi
Maria Jose Caturla
Chu-Chun Fu
Kazuhiro Yasuda
Yanwen Zhang

May 2013

ACKNOWLEDGMENTS

The papers published in this volume result from the 2012 MRS Fall symposium HH. We sincerely thank all of the oral and poster presenters of the symposium who contributed to this proceedings volume. We also thank the reviewers of these manuscripts, who provided valuable feedback to the editors and to the authors. It is an understatement to say that the symposium and the proceedings would not have happened without the organizational help of the Materials Research Society and its staff, particularly the Publications staff for guiding us smoothly through the submission/review process.

MATERIALS RESEARCH SOCIETY SYMPOSIUM PROCEEDINGS

Volume 1477 — Low-Dimensional Bismuth-based Materials, 2012, S. Muhl, R. Serna, A. Zeinert, S. Hirsekor, ISBN 978-1-60511-454-5
Volume 1478 — Nanostructured Carbon Materials for MEMS/NEMS and Nanoelectronics, 2012, A.V. Sumant, A.A. Balandin, S.A. Getty, F. Piazza, ISBN 978-1-60511-455-2
Volume 1479 — Nanostructured Materials and Nanotechnology—2012, 2012, C. Gutiérrez-Wing, J.L. Rodríguez-López, O. Graeve, M. Munoz-Navia, ISBN 978-1-60511-456-9
Volume 1480 — Novel Characterization Methods for Biological Systems, 2012, P.S. Bermudez, J. Majewski, N. Alcantar, A.J. Hurd, ISBN 978-1-60511-457-6
Volume 1481 — Structural and Chemical Characterization of Metals, Alloys and Compounds—2012, 2012, A. Contreras Cuevas, R. Pérez Campos, R. Esparza Muñoz, ISBN 978-1-60511-458-3
Volume 1482 — Photocatalytic and Photoelectrochemical Nanomaterials for Sustainable Energy, 2012, L. Guo, S.S. Mao, G. Lu, ISBN 978-1-60511-459-0
Volume 1483 — New Trends in Polymer Chemistry and Characterization, 2012, L. Fomina, M.P. Carreón Castro, G. Cedillo Valverde, J. Godínez Sánchez, ISBN 978-1-60511-460-6
Volume 1484 — Advances in Computational Materials Science, 2012, E. Martínez Guerra, J.U. Reveles, A. Aguayo González, ISBN 978-1-60511-461-3
Volume 1485 — Advanced Structural Materials—2012, 2012, H. Calderon, H.A. Balmori, A. Salinas, ISBN 978-1-60511-462-0
Volume 1486E — Nanotechnology-enhanced Biomaterials and Biomedical Devices, 2012, L. Yang, M. Su, D. Cortes, Y. Li, ISBN 978-1-60511-463-7
Volume 1487E — Biomaterials for Medical Applications—2012, 2012, S. Rodil, A. Almaguer, K. Anselme, J. Castro, ISBN 978-1-60511-464-4
Volume 1488E — Concrete with Smart Additives and Supplementary Cementitious Materials, 2012, L.E. Rendon Diaz Miron, B. Martinez Sanchez, K. Kovler, N. De Belie, ISBN 978-1-60511-465-1
Volume 1489E — Compliant Energy Sources, 2013, D. Mitlin, ISBN 978-1-60511-466-8
Volume 1490 — Thermoelectric Materials Research and Device Development for Power Conversion and Refrigeration, 2013, G.S. Nolas, Y. Grin, A. Thompson, D. Johnson, ISBN 978-1-60511-467-5
Volume 1491E — Electrocatalysis and Interfacial Electrochemistry for Energy Conversion and Storage, 2013, T.J. Schmidt, V. Stamenkovic, M. Arenz, S. Mitsushima, ISBN 978-1-60511-468-2
Volume 1492 — Materials for Sustainable Development—Challenges and Opportunities, 2013, M-I. Baraton, S. Duclos, L. Espinal, A. King, S.S. Mao, J. Poate, M.M. Poulton, E. Traversa, ISBN 978-1-60511-469-9
Volume 1493 — Photovoltaic Technologies, Devices and Systems Based on Inorganic Materials, Small Organic Molecules and Hybrids, 2013, K.A. Sablon, J. Heier, S.R. Tatavarti, L. Fu, F.A. Nüesch, C.J. Brabec, B. Kippelen, Z. Wang, D.C. Olson, ISBN 978-1-60511-470-5
Volume 1494 — Oxide Semiconductors and Thin Films, 2013, A. Schleife, M. Allen, S.M. Durbin, T. Veal, C.W. Schneider, C.B.Arnold, N. Pryds, ISBN 978-1-60511-471-2
Volume 1495E — Functional Materials for Solid Oxide Fuel Cells, 2013, J.A. Kilner, J. Janek, B. Yildiz, T. Ishihara, ISBN 978-1-60511-472-9
Volume 1496E — Materials Aspects of Advanced Lithium Batteries, 2013, V. Thangadurai, ISBN 978-1-60511-473-6
Volume 1497E — Hierarchically Structured Materials for Energy Conversion and Storage, 2013, P.V. Braun, ISBN 978-1-60511-474-3
Volume 1498 — Biomimetic, Bio-inspired and Self-Assembled Materials for Engineered Surfaces and Applications, 2013, M.L. Oyen, S.R. Peyton, G.E. Stein, ISBN 978-1-60511-475-0
Volume 1499E — Precision Polymer Materials—Fabricating Functional Assemblies, Surfaces, Interfaces and Devices, 2013, C. Hire, ISBN 978-1-60511-476-7
Volume 1500E — Next-Generation Polymer-Based Organic Photovoltaics, 2013, M.D. Barnes, ISBN 978-1-60511-477-4
Volume 1501E — Single-Crystalline Organic and Polymer Semiconductors—Fundamentals and Devices, 2013, S.R. Parkin, ISBN 978-1-60511-478-1
Volume 1502E — Membrane Material Platforms and Concepts for Energy, Environment and Medical Applications, 2013, B. Hinds, F. Fornasiero, P. Miele, M. Kozlov, ISBN 978-1-60511-479-8
Volume 1503E — Colloidal Crystals, Quasicrystals, Assemblies, Jammings and Packings, 2013, S.C. Glotzer, F. Stellacci, A. Tkachenko, ISBN 978-1-60511-480-4
Volume 1504E — Geometry and Topology of Biomolecular and Functional Nanomaterials, 2013, A. Saxena, S. Gupta, R. Lipowsky, S.T. Hyde, ISBN 978-1-60511-481-1

Materials Research Society Symposium Proceedings

Volume 1505E — Carbon Nanomaterials, 2013, J.J. Boeckl, W. Choi, K.K.K. Koziol, Y.H. Lee, W.J. Ready, ISBN 78-1-60511-482-8
Volume 1506E — Combustion Synthesis of Functional Nanomaterials, 2013, R.L. Vander Wal, ISBN 978-1-60511-483-5
Volume 1507E — Oxide Nanoelectronics and Multifunctional Dielectrics, 2013, P. Maksymovych, J.M. Rondinelli, A. Weidenkaff, C-H. Yang, ISBN 978-1-60511-484-2
Volume 1508E — Recent Advances in Optical, Acoustic and Other Emerging Metamaterials, 2013, K. Bertoldi, N. Fang, D. Neshev, R. Oulton, ISBN 978-1-60511-485-9
Volume 1509E — Optically Active Nanostructures, 2013, M. Moskovits, ISBN 978-1-60511-486-6
Volume 1510E — Group IV Semiconductor Nanostructures and Applications, 2013, L. Dal Negro, C. Bonafos, P. Fauchet, S. Fukatsu, T. van Buuren, ISBN 978-1-60511-487-3
Volume 1511E — Diamond Electronics and Biotechnology—Fundamentals to Applications VI, 2013, Y. Zhou, ISBN 978-1-60511-488-0
Volume 1512E — Semiconductor Nanowires—Optical and Electronic Characterization and Applications, 2013, J. Arbiol, P.S. Lee, J. Piqueras, D.J. Sirbuly, ISBN 978-1-60511-489-7
Volume 1513E — Mechanical Behavior of Metallic Nanostructured Materials, 2013, Q.Z. Li, D. Farkas, P.K. Liaw, B. Boyce, J.Wang, ISBN 978-1-60511-490-3
Volume 1514 — Advances in Materials for Nuclear Energy, 2013, C.S. Deo, G. Baldinozzi, M.J. Caturla, C-C. Fu, K. Yasuda, Y. Zhang, ISBN 978-1-60511-491-0
Volume 1515E — Atomic Structure and Chemistry of Domain Interfaces and Grain Boundaries, 2013, S.B. Sinnott, B.P. Uberuaga, E.C. Dickey, R.A. De Souza, ISBN 978-1-60511-492-7
Volume 1516 — Intermetallic-Based Alloys—Science, Technology and Applications, 2013, I. Baker, S. Kumar, M. Heilmaier, K. Yoshimi, ISBN 978-1-60511-493-4
Volume 1517 — Complex Metallic Alloys, 2013, M. Feuerbacher, Y. Ishii, C. Jenks, V. Fournée, ISBN 978-1-60511-494-1
Volume 1518 — Scientific Basis for Nuclear Waste Management XXXVI, 2012, N. Hyatt, K.M. Fox, K. Idemitsu, C. Poinssot, K.R. Whittle, ISBN 978-1-60511-495-8
Volume 1519E — Materials under Extreme Environments, 2013, R.E. Rudd, ISBN 978-1-60511-496-5
Volume 1520E — Structure-Property Relations in Amorphous Solids, 2013, Y. Shi, M.J. Demkowicz, A.L. Greer, D. Louca, ISBN 978-1-60511-497-2
Volume 1521E — Properties, Processing and Applications of Reactive Materials, 2013, E. Dreizin, ISBN 978-1-60511-498-9
Volume 1522E — Frontiers of Chemical Imaging—Integrating Electrons, Photons and Ions, 2013, C.M. Wang, J.Y. Howe, A. Braun, J.G. Zhou, ISBN 978-1-60511-499-6
Volume 1523E — Materials Informatics, 2013, R. Ramprasad, R. Devanathan, C. Breneman, A. Tkatchenko, ISBN 978-1-60511-500-9
Volume 1524E — Advanced Multiscale Materials Simulation—Toward Inverse Materials Computation, 2013, D. Porter, ISBN 978-1-60511-501-6
Volume 1525E — Quantitative *In-Situ* Electron Microscopy, 2013, N.D. Browning, ISBN 978-1-60511-502-3
Volume 1526 — Defects and Microstructure Complexity in Materials, 2013, A. El-Azab, A. Caro, F. Gao, T. Yoshiie, P. Derlet, ISBN 978-1-60511-503-0
Volume 1527E — Scanning Probe Microscopy—Frontiers in Nanotechnology, 2013, M. Rafailovich, ISBN 978-1-60511-504-7
Volume 1528E — Advanced Materials Exploration with Neutrons and Synchrotron X-Rays, 2013, J.D. Brock, ISBN 978-1-60511-505-4
Volume 1529E — Roll-to-Roll Processing of Electronics and Advanced Functionalities, 2013, T. Blaudeck, G. Cho, M.R. Dokmeci, A.B. Kaul, M.D. Poliks, ISBN 978-1-60511-506-1
Volume 1530E — Materials and Concepts for Biomedical Sensing, 2013, P. Kiesel, M. Zillmann, H. Schmidt, B. Hutchison, ISBN 978-1-60511-507-8
Volume 1531E — Low-Voltage Electron Microscopy and Spectroscopy for Materials Characterization, 2013, R.F. Egerton, ISBN 978-1-60511-508-5
Volume 1532E — Communicating Social Relevancy in Materials Science and Engineering Education, 2013, K. Chen, R. Nanjundaswamy, A. Ramirez, ISBN 978-1-60511-509-2
Volume 1533E — The Business of Nanotechnology IV, 2013, L. Merhari, D. Cruikshank, K. Derbyshire, J. Wang, ISBN 978-1-60511-510-8

Materials Research Society Symposium Proceedings

Volume 1534E — Low-Dimensional Semiconductor Structures, 2012, T. Torchyn, Y. Vorobie, Z. Horvath, ISBN 978-1-60511-511-5

Prior Materials Research Society Symposium Proceedings available by contacting Materials Research Society

Theory, Modeling and Simulation

Mater. Res. Soc. Symp. Proc. Vol. 1514 © 2013 Materials Research Society
DOI: 10.1557/opl.2013.43

Multiscale Approach to Theoretical Simulations of Materials for Nuclear Energy Applications: Fe-Cr and Zr-based Alloys

Igor A. Abrikosov[1], Alena V. Ponomareva[2], Svetlana A. Barannikova[3,4], Olle Hellman[1], Olga Yu. Vekilova[1], Sergei I. Simak[1] and Andrei V. Ruban[5]
[1]Department of Physics, Chemistry and Biology (IFM), Linköping University, SE-581 83 Linköping, Sweden.
[2]Theoretical Physics and Quantum Technology Department, National University of Science and Technology "MISIS", RU-119049 Moscow, Russia.
[3]Institute of Strength Physics and Materials Science, Siberian Branch of Russian Academy of Science, Akademicheskii Pr. 2/4, 634021 Tomsk, Russia.
[4]Department of Physics and Engineering, Tomsk State University, 36 Lenin Prospekt, 634050 Tomsk, Russia.
[5]Applied Material Physics, Department of Materials Science and Engineering, Royal Institute of Technology (KTH), SE-100 44 Stockholm, Sweden

ABSTRACT

We review basic ideas behind state-of-the-art techniques for first-principles theoretical simulations of the phase stabilities and properties of alloys. We concentrate on methods that allow for an efficient treatment of compositional and thermal disorder effects. In particular, we present novel approach to evaluate free energy for strongly anharmonic systems. Theoretical tools are then employed in studies of two materials systems relevant for nuclear energy applications: Fe-Cr and Zr-based alloys. In particular, we investigate the effect of hydrostatic pressure and multicomponent alloying on the mixing enthalpy of Fe-Cr alloys, and show that in the ferromagnetic state both of them reduce the alloy stability at low Cr concentration. For Zr-Nb alloys, we demonstrate how microscopic parameters calculated from first-principles can be used in higher-level models.

INTRODUCTION

Nuclear energy has become an important part of the energy portfolio for the modern society, contributing, e.g. to reduction of the dependence on the fossil fuels and emission of green-house gases. Advances of the technology would not be possible without tremendous work invested in design of materials, operating for extended periods of time at high temperatures, under irradiation, stress and corrosion. Moreover ongoing development of next generation reactors strengthens demands on materials to be used in fission and future fusion reactors, which include good tensile and creep strength, as high as possible operational temperatures, a control over ductile to brittle transition temperature, resistance to irradiation, high thermal conductivity, low residual activation, compatibility with cooling media, and good weldability [1]. A great challenge is to identify potentially significant materials, to develop efficient technological processes, and to optimize their functionality. Solution to these tasks clearly requires increasing understanding of nuclear materials performance under extreme conditions. In this respect, a new and powerful instrument is now available for researches, computers. Their advances initiated a development of a qualitatively new brunch in science, computer modeling. The major goal of theory in applications within the materials sciences is to provide fundamental understanding of materials properties and behavior at different external conditions, to accelerate the design of new materials and to discover phenomena with high potential to improve the existing technologies, as well as to create qualitatively new technologies. Therefore, the development of the basic theory, computational algorithms and qualitative models is an important task within the field of materials modeling.

In this work we briefly review recent developments in the field of *ab initio* electronic structure theory and its applications for studies of complex alloy systems. Basic ideas behind state-of-the-art techniques for first-principles theoretical simulations of the phase stabilities and properties of intermetallic compounds and alloys based on multiscale approach are outlined. We concentrate on methods that allow for an efficient treatment of compositional [2] and thermal [3] disorder effects, and illustrate their performance for two systems relevant for nuclear energy applications: Fe-Cr and Zr-based alloys.

In particular, ferritic steels are used to manufacture reactor pressure vessels (RPV). Irradiation-induced accelerated ageing is considered to be a crucial issue that limits the lifetime of nuclear reactors. Fe-Cr steels with 7-18 at.% Cr are considered as promising structural materials for fast neutron reactors due to their relatively low rate of swelling at elevated temperatures [1]. In the binary Fe-Cr alloy a spinodal decomposition can lead to a formation of precipitates of α' phase, but at low chromium concentrations the alloys are anomalously stable [4]. The effect has been a subject of active theoretical investigations [5-7], which significantly improved the understanding of the binary Fe-Cr system. On the other hand, in reality promising steels for RPV contain other elements, including significant amounts of Ni, Mn, and Mo. Their effect on the stability of parent Fe-Cr alloy did not received corresponding attention of the theory. Moreover, the radiation-enhanced formation of defects, like interstitial clusters, dislocations, etc., may lead to the presence of highly compressed areas within the material. How would the local pressure affect the stability of the alloy? Here we investigate the effect of hydrostatic pressure and multicomponent alloying on the mixing enthalpy of Fe-Cr alloys, and show that in the ferromagnetic state both of them reduce the alloy stability at low Cr concentrations. We analyze the effect of pressure in terms of the effective cluster interactions [2], which are *ab initio* parameters for higher level modeling within the multiscale approach.

Hexagonal closed packed (hcp)Zr-based alloys represent another material system widely used in nuclear energy applications, e.g. as fuel cladding materials. Alloying has turned out to provide a significant improvement of materials properties as compared to pure Zr. In particular, Zr-Nb alloys (with about 1 at.% Nb) show strengthening behavior due to polymorphous martensitic α↔β transition. However, an empirical approach is still taken for the selection of alloying additions [1], and improved understanding of Zr-based alloys is clearly required. Here we demonstrate how microscopic parameters calculated from first-principles can be used in higher-level phenomenological models to correlate elastic and plastic behavior of alloys. We argue that the phase transition in Zr-Nb system is associated with pronounced dynamical instabilities of body-centered phase of Zr, and present novel approach to evaluate free energy and describe phase transitions in strongly anharmonic systems [3].

THEORY

In the field of materials modeling a consensus has emerged that a successful theoretical description of materials may be achieved most efficiently in the framework of the so-called multiscale modeling. By this one understands a solution of the complete simulation problem step-by-step employing theoretical methodologies, which are suitable for the particular length and/or time scale followed by an appropriate coarse graining when proceeding towards the next (larger) scale. For example, one starts with the solution of the quantum mechanical problem within density functional theory (DFT) for a relatively small system (~100 atoms). From these results one determines interactions between different atoms and uses them in simulations (classical molecular dynamics or statistical mechanics, e.g. Monte-Carlo technique), which includes 10^3 to 10^5 particles. In doing so one can study, for example, crack propagations, melting, spinodal decomposition or ordering in metals and alloys. For simulations of mechanical properties and microstructure (i.e. properties on the scale of microns) one often needs to substitute the atomistic description by continuum theories. The latter can be done, for example, in the framework of phenomenological models.

The multiscale modeling experienced certain success. But the fact is that at present its stages are disconnected to a large degree. DFT calculations treat mostly the electronic subsystem, while atomic motions are often neglected. In classical molecular dynamics the lack of input experimental information is often compensated by oversimplified interpolations. Even more this is true for the applications of continuum models at large time and length scales. Unfortunately, a creation of a consistent scheme for the multiscale modeling of materials properties, where all stages are connected to each other, is still a challenge. Below we briefly review recent advances in theoretical treatment of compositional and thermal disorder.

Total energy calculations at zero temperature

Starting with the lowest level description of a simulation system at the DFT level at temperature T=0K, a solution to the quantum mechanical problem for a selected (fixed) atomic configuration *{R}* is provided by the electronic structure theory [2]. Within the DFT one solves the so-called Kohn-Sham (KS) equations for a system of independent "electrons" (characterized by single-particle wave functions). We denote the "electrons" density as $n(r,\{R\})$ and note that it depends on the atomic configuration *{R}* and coordinate r (in contrast to many body wave function, which depends on the coordinates of *all* the electrons in the system). The essence of the KS theory is that it is constructed in such a way that the density of the independent "electrons" obtained from a self-consistent solution of the KS equations is exactly the same as the electron density in the original system of real interacting electrons. The DFT theorems ensure that the converged solution of the KS equations corresponds to the charge density in the (electronic) ground state. The total (potential) energy of a configuration *{R}* can now be calculated as a functional of the one-electron density:

$$E_{\{R\}}^{DFT}[n] = T_s[n] + \int d^3r V_{ext}(r,\{R\}) n(r,\{R\}) + \iint d^3r d^3r' \frac{n(r,\{R\}) n(r',\{R\})}{|r-r'|} + E_{II} + E_{xc}[n] \quad (1)$$

In Eq. (1) $V_{ext}(r,\{R\})$ is the external potential (e.g. due to nuclei), E_{II} is the classical electrostatic interaction between the nuclei,

$$T_s[n] = \sum_{occ} \varepsilon_{one-el} - \int d^3r V_{KS}(r,\{R\}) n(r,\{R\}) \quad (2)$$

is the kinetic energy of independent electrons with eigenstate energies $\varepsilon_{one\text{-}el}$ moving in the effective potential $V_{KS}(r,\{R\})$. These quantities are obtained by the self-consistent solution of the KS equations, together with the charge density n. The last term in Eq. (1) is the so-called exchange-correlation energy $E_{xc}[n]$, which incorporates the effects of many-body electron-electron interactions, and which must be approximated. For simulations at T=0K presented in this work we used the approximation advocated in Ref. [2], that is we calculated the self-consistent electron densities within the local-density approximation (LDA) [8] and then the total energies in Eq. (1) were calculated in the generalized-gradient approximation (GGA) [9].

A serious problem with the application of the DFT formalism occurs if a configuration $\{R\}$ does not have any translational periodicity, like for a solid solution phase $A_{1-x}B_x$. The point is that in this case group theory, which is the corner stone of the modern electronic structure calculations, cannot be used directly. One obvious way to deal with a disordered system is to consider its fragment(s), to impose periodic boundary conditions, and to solve the DFT problem for such "supercells". In many cases this way of simulating the disorder is acceptable for total energy calculations. However, it is important to remember that this model is based on the use of translational symmetry, which is absent in real random alloys. Moreover, the choice of a supercell is not a trivial task. The problem is that whether a supercell can be considered as (quasi-)random or not in the total energy calculations is exclusively dictated by the nature of the interatomic interactions in this particular system. Careful discussion of the problem can be found in Ref. [2].

As an alternative to the supercell approach one can reconstruct three-dimensional periodicity of the solid solution phase $A_{1-x}B_x$ by mapping it onto an ordered lattice of "effective" atoms, which describe the original system on the average. In terms of the electronic structure problem, one is ultimately interested in processes of electron scattering off the atoms in the system, the so-called multiple-scattering. For completely disordered alloys very successful approximation, the coherent potential approximation (CPA), is constructed by placing effective scatterers at the sites of the original system. Scattering properties of these effective atoms have to be determined self-consistently from the condition that the scattering of electrons off the alloy components, embedded in the effective medium as impurities, vanishes on the average [10]. The CPA is currently one of the most popular techniques to deal with substitutional disorder, and in this work we use its implementation within the exact muffin-tin orbital (EMTO) theory [11]. Details of the calculations for Fe-Cr alloys are the same as in Ref. 7, while for Zr-Nb alloys they are summarized in Ref. [12]

Free energy calculations at finite temperature

In order to describe phase equilibria and to construct phase diagrams within the multiscale approach, we have to move to higher-level theories. One possibility here is offered within the phenomenological thermodynamics approach. For alloy systems one considers most often temperature-composition phase diagrams at ambient pressure. Thus, one deals with a system in thermal and mechanical contact with a constant-temperature constant-pressure heat bath, whose equilibrium is described by the thermodynamic potential

$$G=E+PV-TS=H-TS=F+PV, \quad (3)$$

where G is the Gibbs free energy, E is the energy of the system, P, V, T represent pressure, volume, and temperature, respectively, S denotes the entropy, $H= E+PV$ is the enthalpy, and $F= E-TS$ is the Helmholtz free energy. If phase α transforms into phase β in a pure material at equilibrium transformation temperature T_c, $G_\alpha = G_\beta$ at T_c.

In a binary system $A_{1-x}B_x$ at constant P and T the G curves as a function of an alloy component fraction x for the phases in equilibrium must share a common tangent. However, this condition defines areas at temperature-composition phase diagram subject to a transformation via nucleation-and-growth mechanism, like the binodal decomposition in Fe-Cr alloys. The alloys within this area are metastable. A condition for the spinodal decomposition, which occurs in systems unstable to any concentration fluctuation, is given by inequality $dG/dx<0$. The simplest calculations here can be carried out within the mean-filed approximation [2]. One assumes a complete disorder among the alloy components for solution phases and calculates their enthalpies $H_{A_{1-x}B_x}$ from DFT total energies and pressure, obtained e.g. by means of the supercell approach or within the CPA, as described in the previous section. Then one defines the mixing enthalpy ΔH^{mix} and configurational contribution to the mixing entropy ΔS^{mix} as

$$\Delta H^{mix}_{A_{1-x}B_x} = H_{A_{1-x}B_x} - (1-x)H_A - xH_B \qquad \Delta S^{mix}_{A_{1-x}B_x} = -k_B\left[x\ln x + (1-x)\ln(1-x)\right], \tag{4}$$

where $H_{A(B)}$ represents the enthalpy of a standard state, for simplicity taking here as pure alloy components, k_B is the Boltzmann constant. It is now possible to proceed with a construction of the phase diagram. This highly oversimplified approach is supposed to overestimate the order-disorder transition temperature, because of the neglect of the short-range order effects in solid solutions. But in systems with a tendency towards phase separation it may give surprisingly good agreement with more advanced calculations [13], and analysis of the mixing enthalpy often provides important insight into behavior of alloys upon ordering of phase separation. This will be illustrated further in section Discussion.

To go beyond the mean-filed description, one most often uses the so-called canonical ensemble. The thermodynamic potential is the Helmholtz free energy F, which describes a system at constant temperature and volume, and which is calculated as:

$$F(T,V,N) = -k_B T \ln Z(T,V,N) = -k_B T \ln\left[\sum_{\{R\}} \exp\left(-\frac{F_{\{R\}}}{k_B T}\right)\right] \tag{5}$$

where $Z(T,V,N)$ is the partition function and k_B is the Boltzmann constant. Because for the problem of configurational thermodynamics, which deals with diffusion time scales, the fast degrees of freedom, such as electronic, vibrational and magnetic, can be averaged out, the partition function for the configurational degrees of freedom in Eq. (5) should consist of the free energies $F_{\{R\}}$ of electronic, magnetic and vibrational excitations of alloys with a given configuration $\{R\}$ [2]. In practice DFT calculations are almost always carried out at zero temperature. In principle, vibrational contribution F_{vib} can be included within the so-called quasi-harmonic approximation [2,14] (QHA), where it is calculated from the phonon dispersion relations via the phonon density of states $g(\omega)$ using expression:

$$F_{vib} = \int_0^{\infty} g(\omega)\left[k_B T \ln\left(1 - \exp\left(\frac{h\omega}{k_B T}\right)\right) + \frac{h\omega}{2}\right] d\omega \tag{6}$$

However, for large class of alloy systems this will not work. For example, phonon dispersion relations calculated for the body-centered cubic (bcc) phase of Zr at T=0K show imaginary frequencies [3]. This means that free energy cannot be defined within the QHA. On the other hand, bcc Zr is stabilized at high temperature by the anharmonic effects, and bcc (β-phase) precipitates are very important for the improvement of mechanical properties of Zr-Nb alloys, which will be discussed below.

An accurate and easily extendable method to deal with lattice dynamics of solids, the Temperature Dependent Effective Potential Method (TDEP) has been suggested in [3]. It is designed to work even for strongly anharmonic systems where the traditional quasi-harmonic approximation fails. The method is based on *ab initio* molecular dynamics (AIMD) simulations and provides a consistent way to extract the best possible harmonic—or higher order—potential energy surface at finite temperatures. Using a Taylor expansion of the potential energy U in time dependent atomic displacements $u(t)$, truncated after the second-order terms:

$$U = U_0 + \frac{1}{2}\sum_{ij\mu\nu} u_{j\nu}(t)\ D^{ij}_{\mu\nu}\ u_{i\mu}(t), \tag{7}$$

where $D^{ij}_{\mu\nu}$ is the force constant matrix, i and j denote atomic positions, and μ,η are the Cartesian coordinates, TDEP method exploits the potential energy surface at *finite* temperature T by means of AIMD, generating a set of displacements $\boldsymbol{u}_i(t)$ and forces $\boldsymbol{F}_i(t)$, which are consistent with each other. The TDEP force constant matrix can now be determined at the temperature of MD simulation by minimizing the difference between AIMD forces and TDEP forces $\boldsymbol{F}_i(t) = \sum_j \boldsymbol{D}^{ij}\boldsymbol{u}_j(t)$ for the entire MD run. As has been demonstrated in Ref. [3], the procedure is unique, well converged, and at finite temperatures it eliminates the imaginary frequencies obtained in QHA calculations at T=0, bringing the phonon dispersion relations for bcc Zr in very good agreement with available experiment. Consequently, the free energy can now be defined as $F_{TDEP} = U_0 + F_{vib}$, where F_{vib} is given by Eq. (6) and U_0 is determined from AIMD and Eq. (7) as

$$U_0 = \left\langle U_{MD}(t) - \sum_{ij\mu\nu}\frac{1}{2} D^{ij}_{\mu\nu} u_{j\nu}(t) u_{i\mu}(t)\right\rangle \tag{8}$$

It is important to understand that in Eq. (5) the sum runs over all possible states $\{R\}$ of the system. For example, if we need to find F and/or Z for $A_{1-x}B_x$ alloy at fixed composition x and underlying bcc crystal lattice, $\{R\}$ represents all possible occupations of the bcc lattice by $(1-x)N$ A and xN B atoms. The basic idea of the formalism, which represents a typical first step of the multiscale modeling, is to coarse-grain the problem by eliminating quantum degrees of freedom from the Hamiltonian. This is achieved by mapping DFT total energy functional, Eq. (1), on the configurational energies calculated from classical generalized Ising Hamiltonian:

$$H^{conf} = \frac{1}{2}\sum_s V_s^{(2)} \sum_{i,j\in s} c_i c_j + \ldots \tag{9}$$

In Eq. (9) i,j are lattice sites, the occupation numbers c_i in case of a binary alloy take on values 1 or 0 depending on the type of atom occupying site i, and $V_s^{(d)}$ are the effective cluster interactions (ECI), which correspond to clusters of the order d and type s. Different methods have been developed for calculations of ECI from first principles [2]. In this work we employed the so-called screened generalized perturbation method (SGPM) [15].

DISCUSSION

Fe-Cr alloys

A moderate amount of Cr, up to 10%, in ferritic steels has proven to be most beneficial to their ductile to brittle transition temperature as well as to their corrosion resistance and resistance to neutron radiation induced swelling. Though there is a miscibility gap for the alloy concentrations above 10 at. % Cr at room temperature for lower concentrations of Cr the alloy is stable with respect to this decomposition. Moreover, diffuse-neutron-scattering experiments show an anomaly associated with a change of the ordering tendency from the clustering in Cr-rich alloys towards the short-range order for low Cr Fe-Cr alloys, and this effect is now well understood form *ab initio* theory [4-6]. In this work we investigate how this tendency is affected by pressure, which may be present, e.g. locally due to radiation induced defects, as well as by other alloying elements, present in RPV steels.

Figure 1 shows the calculated mixing enthalpy and its second derivative of ferromagnetic Fe-Cr alloys as a function of Cr concentration at ambient pressure and at 20 GPa. The standard states of Fe H_{Fe} in Eq. (4) is the ferromagnetic bcc Fe, while the reference state of bcc Cr is

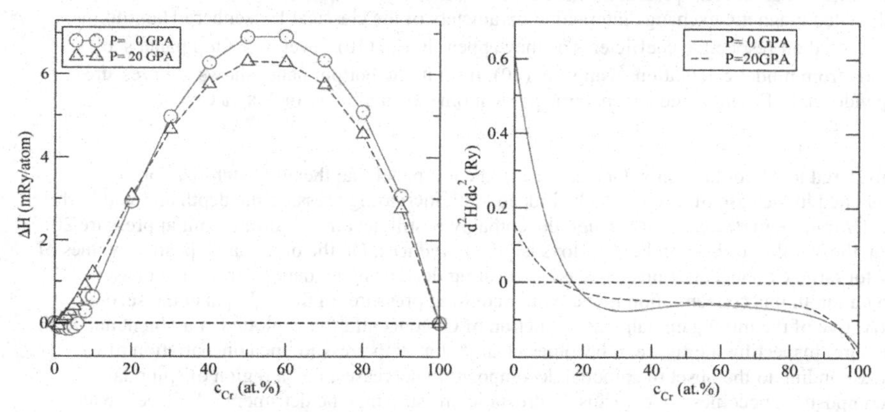

Figure 1. (left) Calculated mixing enthalpy ΔH and (right) its second derivative of ferromagnetic Fe-Cr alloys as a function of Cr concentration at pressures $P = 0$ (full red line, circles) and 20 GPa (dashed blue line, diamonds).

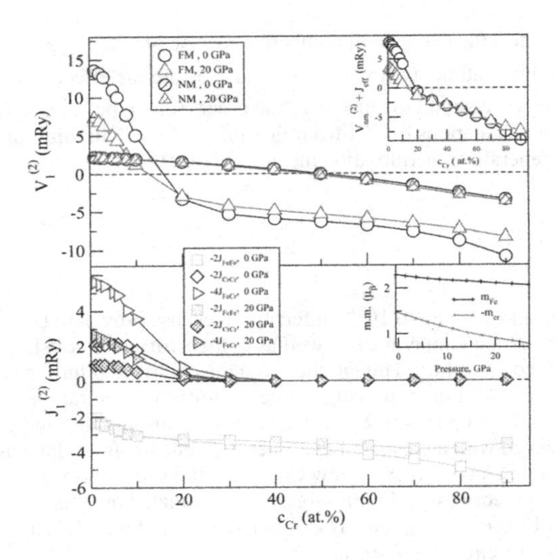

Figure 2. (top) Calculated pair ECI between nearest neighbors $V_1^{(2)}$, Eq. (9), for ferromagnetic (open symbols) and nonmagnetic (dashed symbols) Fe-Cr alloys as a function of Cr concentration at ambient pressure (black circles) and at P=20 GPa (red trangles); (bottom) Calculated magnetic exchange interaction parameters of the classical Heisenberg Hamiltonian J_{AB} with their respective coefficients as they appear in Eq. (10). Inset in the top panel shows results from model calculations using Eq. (10). Inset in the bottom panel shows the pressure dependence of Fe (m_{Fe}) and Cr (m_{Cr}) magnetic moments in alloy with 3 at.% Cr.

considered to be nonmagnetic. One can see in the left panel that the alloy stabilization is weakened in the case of low-Cr steels. Indeed, with increasing pressure the depth and width of the ΔH minimum decreases. Although the enthalpy is still negative in dilute limit at pressure 20 GPa, the stability of Fe-rich Fe-Cr alloys is greatly reduces. On the other hand, positive values of ΔH for large Cr concentrations decrease, indicating decreasing tendency toward the phase separation in this concentration range with increasing pressure. In the right panel the second derivative of the mixing enthalpy as a function of Cr concentration is plotted at ambient and high pressure. Inspecting the figure, it becomes clear that at high pressure the concentration of Cr corresponding to the onset of spinodal decomposition decreases, i.e., a region of spinodal decomposition becomes wider. Thus, hydrostatic pressure may be detrimental for steels with high Cr composition, close to the border of the spinodal region on the phase diagram.

To understand this behavior, we show in Fig. 2 calculated pair ECI between nearest neighbors $V_1^{(2)}$, Eq. (9), as a function of Cr concentration and pressure. Its negative sign for most of the compositions indicates tendency towards clustering, but in Fe-rich alloys the sign is positive, and the alloys are stable. However, the increase of pressure reduces $V_1^{(2)}$ at low Cr concentration, indicating the alloy destabilization, in agreement with conclusion drawn from ΔH calculations. Because in low Cr steels magnetic moments of Cr are antiparallel to magnetic moments of Fe in the (magnetic) ground state [4-7], one can approximate the ECI as a sum of chemical and magnetic terms following a model of Ref. [7] as

$$V_s^{(2,\mathrm{mod})} = V_s^{(2,chem)} + V_s^{(2,magn)} \qquad V_s^{(2,magn)} = -2(J_{FeFe} + 2J_{FeCr} + J_{CrCr}) \tag{10}$$

where J_{AB} is the magnetic exchange interaction parameter of the classical Heisenberg Hamiltonian. Assuming further that $V_s^{(2,chem)}$ does not depend on the magnetic state, it can be associated with ECI in nonmagnetic alloys. In Fig. 2 it is clearly seen that this term depends weakly on alloy concentration, and it is quite small. Thus, most important contribution into ECI and therefore the stability trends of ferromagnetic Fe-Cr alloys comes explicitly from the magnetic exchange interactions. In Fe-rich alloys at ambient pressure $J_{CrCr} \approx -J_{FeFe}$, and therefore $V_s^{(2,chem)}$ is dominated by J_{FeCr}. However, with increasing pressure magnetic moment at Cr is reduced much stronger than at Fe, as shown in the bottom panel inset in Fig. 3. Thus, J_{CrCr} and J_{FeCr} are reduced much stronger in magnitude than J_{FeFe}, making $V_s^{(2,chem)}$, and consequently $V_1^{(2)}$ less positive. The example above demonstrates how the parameters of higher-level models obtained from first-principles can be used for better understanding of materials behavior. Of course, they can be used directly for modeling of phase diagrams [16].

To investigate the effect of additional alloying elements on the stability of bcc Fe-Cr steels, we calculate the mixing enthalpies and their second derivatives for Fe-Cr-Ni and Fe-Cr-Ni-Mn-Mo systems. The results are shown in Fig. 3. One can see that these elements tend to

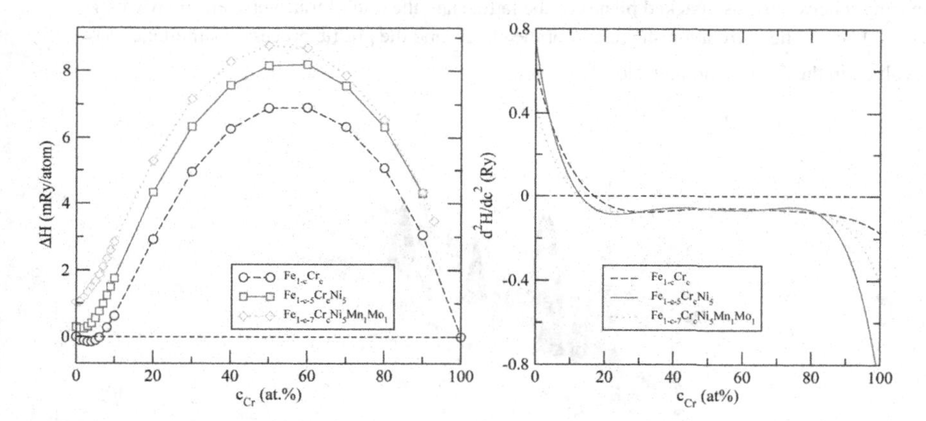

Figure 3. (left) Calculated mixing enthalpy and (right) its second derivative of ferromagnetic binary Fe-Cr alloys (dashed blue line, circles), as well as ternary alloys with 5 at. % Ni (full red line, squares), and multicomponent alloys with 5 at. % Ni, 1 at. % Mn and 1 at.% Mo (dotted green line, diamonds) as a function of Cr concentration

destabilize solid solutions and enhance the tendency towards the spinodal decomposition. Indeed, ΔH is now positive for all Cr compositions, and d^2H/dc^2 changes sign at lower Cr concentrations. More detailed analysis of multicomponent alloying of Fe-Cr alloys will be presented elsewhere.

Zr-Nb alloys

Relatively low mechanical strength of pure Zr is a well-known problem for its applications as a construction material. For nuclear energy applications it is alloyed, e.g. with Nb, which leads to strengthening behavior. Understanding mechanical properties of this alloy is, therefore, of primary importance. In Fig. 4 we show typical localization pattern observed in experimental studies of plastic flow of Zr alloyed with 1at.%Nb by mechanical testing using a technique of double-exposure speckle-photography [17]. It unambiguously demonstrates a tendency towards localization behavior from yield point to failure. The autowave features of the localized plastic flow patterns are of major importance at the linear stage of deformation hardening as the plastic flow localization takes on the form of phase autowave, which has length $\lambda \approx 10^{-2}$ m and propagation rate $10^{-5} \leq V_{aw} \leq 10^{-4}$ m·s^{-1}. Unfortunately, at present direct simulations at such length and time scales are beyond the capability of *ab initio* theory. However, within the multiscale approach it can give useful information for higher-level theories and/or phenomenological models, supplying reliable parameters for the latter which are difficult or impossible to measure experimentally.

For example, a model of plastic flow localization introduced in [18,19] suggested that there is a correlation between macroscopic parameters of the autowaves, λ and V_{aw} for easy glide and linear work hardening stages to the materials specific microscopic parameters, the spacing d between close-packed planes of the lattice and the rate of transverse elastic waves $V_{\perp}$: $\lambda V_{aw} \approx \frac{1}{2} d V_{\perp}$. The correlation suggests that the elastic and the plastic processes simultaneously involved in the deformation are closely related.

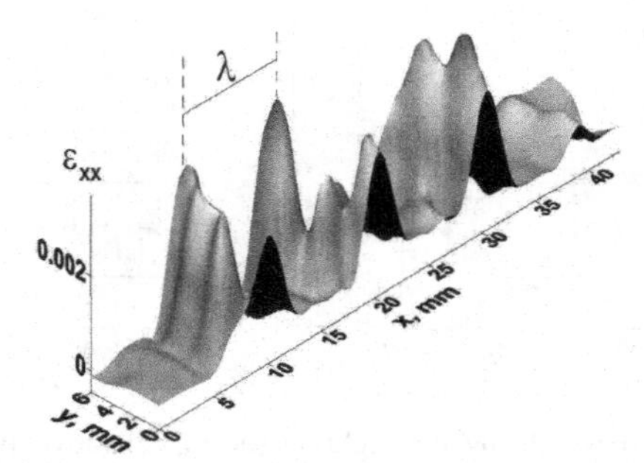

Figure 4. Localized plastic flow autowave generated at the linear work hardening stage of Zr alloyed with 1at.%Nb; ε_{xx} - local elongation; x and y- specimen length and width, respectively; λ - nucleus spacing (autowave length).

The reliability of this model was demonstrated for a set of pure metals, but for metallic alloys the experimental information on the microscopic parameters was often absent in the literature. By means of EMTO-CPA method (see section Theory) one can calculate equilibrium volume V per atom and bulk modulus B. In particular, for Zr-Nb alloy, shown in Fig. 4, we considered hcp crystal structure and obtained V=23.4 Å^3 with c/a ratio 1.62 and B=90 GPa, very close to corresponding experimental values for pure Zr. These values can be recalculated [12] into d=2.46 Å and $V_\perp$=2400 m/s, giving $2\lambda V_{aw}/dV_\perp \approx 0.77$. In Ref. [12] it was demonstrated that condition $2\lambda V_{aw}/dV_\perp \approx 1$ is fulfilled even better for other alloy systems. One possibility here could be that we assumed hcp crystal structure for the Zr-Nb alloy, while it is believed that its improved strengthening behavior as compared to pure Zr is due to the presence of bcc-phase precipitates. The latter should affect the plastic flow patterns. However, simulations of phase stability for Zr-Nb system represent particularly challenging task because of dynamical instability of bcc Zr, discussed above. Thus, development of new theoretical tools for calculations of alloy thermodynamic and mechanical properties in strongly anharmonic solids, e.g. employing the TDEP method for free energy calculations (see section Theory) represents an important and timely task.

CONCLUSIONS

Recent developments in the field of *ab initio* electronic structure theory and its application for studies of complex alloy systems have been reviewed. State-of-the art techniques for calculations of total energy at zero temperature in the framework of Density Functional Theory and free energy at finite temperature within the multiscale approach have been presented. Also, we have described a novel technique, an accurate and easily extendable method to deal with lattice dynamics of strongly anharmonic solids, the Temperature Dependent Effective Potential Method (TDEP). Considering Fe-Cr alloys, we have investigated the effect of hydrostatic pressure and multicomponent alloying on their phase stability. We have shown that it reduces the stability of ferromagnetic alloys at low Cr concentration and vice versa, makes the solid solution more stable at higher concentrations. To analyze this effect, we have used a model in which the effective pair interactions are split into chemical and magnetic terms, and have shown that the effect of pressure on the phase stability in this system comes mostly through the decrease of the magnetic exchange interactions between Fe-Cr and Cr-Cr pairs induced by a decrease of Cr magnetic moment. Also, we have shown that alloying elements Ni, Mn, and Mo, present in RPV steels, reduce the stability of low-Cr steels against binodal, as well as spinodal decomposition. Considering Zr-Nb alloys, we demonstrated a possibility of obtaining relevant parameters of higher-level phenomenological models from *ab initio* electronic structure calculations, underlying the importance of the multiscale approach.

ACKNOWLEDGMENTS

This study was supported in part by the Ministry of Education and Science of the Russian Federation within the framework of Program "Research and Pedagogical Personnel for Innovative Russia (2009-2013) " (project no. 14.B37.21.0890 of 10.09.2012) . Support from the Swedish Research Council (VR) projects 621-2008-5535, 621-2009-3619, and LiLi-NFM, the

Swedish Foundation for Strategic Research (SSF) program SRL10-0026, Knut & Alice Wallenberg Foundation (KAW) project "Isotopic Control for Ultimate Material Properties", Russian Foundation for Basic Researches (Grant No. 10-02-00-194a, A.V.P.) , and European Research Council (Grant No. 228074, ALPAM) is gratefully acknowledged. SSI acknowledges support from AFM research environment at LiU. Supercomputer resources were provided by the Swedish National Infrastructure for Computing (SNIC) and the Joint Supercomputer Center of RAS (Moscow).

REFERENCES

1. E. A. Marquis, J. M. Hyde, D. W. Saxey, S. Lozano-Perez, V. de Castro, D. Hudson,C. A. Williams, S. Humphry-Baker and G. D.W. Smith, Materials Today **12**, 30 (2009)
2. A. V. Ruban and I. A. Abrikosov, Rep. Prog. Phys. **71**, 046501 (2008).
3. O. Hellman, I. A. Abrikosov, and S. I. Simak, Phys. Rev. B **84**, 180301(R) (2011).
4. P. Olsson, I. A. Abrikosov, L. Vitos, and J. Wallenius, J. Nucl. Mat. **321**, 84 (2003); P. Olsson, I. A. Abrikosov, and J. Wallenius, Phys. Rev. B **73**, 104416 (2006).
5. T. P. C. Klaver, R. Drautz, and M. W. Finnis, Phys. Rev. B **74**, 094435 (2006); M. Yu. Lavrentiev, R. Drautz, D. Nguyen-Manh, T. P. C. Klaver, and S. L. Dudarev, *ibid.* **75**, 014208 (2007); P. A. Korzhavyi, A. V. Ruban, J. Odqvist, J.-O. Nilsson, and B. Johansson, ibid., **79**, 054202 (2009).
6. A. V. Ruban, P. A. Korzhavyi, and B. Johansson, Phys. Rev. B **77**, 094436 (2008).
7. A. V. Ponomareva, A. V. Ruban, O. Yu. Vekilova, S. I. Simak, and I. A. Abrikosov, Phys. Rev. B **84**, 094422 (2011).
8. J. P. Perdew and Y. Wang, Phys. Rev. B **45**, 13244 (1992).
9. Y. Wang and J. P. Perdew, Phys. Rev. B **44**, 13298 (1991); J. P. Perdew, J. A. Chevary, S. H. Vosko, K. A. Jackson, M. R. Pederson, D. J. Singh, and C. Fiolhais, *ibid.* **46**, 6671 (1992).
10. P. Soven, Phys. Rev. **156**, 809 (1967).
11. L. Vitos, I. A. Abrikosov, and B. Johansson, Phys. Rev. Lett. **87**, 156401 (2001).
12. A. Barannikova, A. V. Ponomareva, L. B. Zuev, Yu. Kh. Vekilov, and I. A. Abrikosov, Solid State Commun. **152**, 784 (2012).
13. B. Alling, A. V. Ruban, A. Karimi, L. Hultman, and I. A. Abrikosov, Phys. Rev. B **83**, 104203 (2011).
14. A. van de Walle and G. Ceder, Rev. Mod. Phys **74**, 11 (2002); G. Grimvall, B. Magyari-Köpe, V. Ozolins, K. A. Persson, *ibid.* **84**, 945 (2012).
15. A. V. Ruban and H. L. Skriver, Phys. Rev. B **66**, 024201 (2002); A. V. Ruban, S. I. Simak, P. A. Korzhavyi, and H. L. Skriver, *ibid.* **66**, 024202 (2002); A. V. Ruban, S. Shallcross, S. I. Simak, and H. L. Skriver, *ibid.* **70**, 125115 (2004).
16. A. V. Ruban and V. I. Razumovskiy, Phys. Rev. B (in press)
17. L.B. Zuev, V.V. Gorbatenko, S.N.Polyakov, "Instrumentation for speckle interferometry and techniques for investigating deformation and fracture", *Proceedings of SPIE,* Vol. 4900, Part 2, pp. 1197-1208 (2002).
18. L.B. Zuev and S.A. Barannikova, Natural Science **2,** 476 (2010).
19. L.B. Zuev and S.A. Barannikova, J. Mod. Phys. **1**, 1-8 (2010).

Mater. Res. Soc. Symp. Proc. Vol. 1514 © 2013 Materials Research Society
DOI: 10.1557/opl.2013.197

Simulating Radiation-Induced Defect Formation in Pyrochlores

David S.D. Gunn[1], John A. Purton[1] and Ilian T. Todorov[1]
[1]Scientific Computing Department, Science & Technology Facilities Council, Daresbury Laboratory, Sci-Tech Daresbury, Keckwick Lane, Daresbury, WA4 4AD, U.K.

ABSTRACT

The accuracy and robustness of new Buckingham potentials for the pyrochlores $Gd_2Ti_2O_7$ and $Gd_2Zr_2O_7$ is demonstrated by calculating and comparing values for a selection of point defects with those calculated using a selection of other published potentials and our own *ab inito* values. Frenkel pair defect formation energies are substantially lowered in the presence of a small amount of local cation disorder. The activation energy for oxygen vacancy migration between adjacent O_{48f} sites is calculated for Ti and Zr pyrochlores with the energy found to be lower for the non-defective Ti than for the Zr pyrochlore by ~0.1 eV. The effect of local cation disorder on the $VO_{48f} \rightarrow VO_{48f}$ migration energy is minimal for $Gd_2Ti_2O_7$, while the migration energy is lowered typically by ~43 % for $Gd_2Zr_2O_7$. As the healing mechanisms of these pyrochlores are likely to rely upon the availability of oxygen vacancies, the healing of a defective Zr pyrochlore is predicted to be faster than for the equivalent Ti pyrochlore.

INTRODUCTION

The disposal and safe storage of nuclear waste is a significant challenge for the global community. Several of the radionuclides generated through the nuclear fuel cycle, such as ^{239}Pu and ^{235}U, have long half lives (24,100 years and 7×10^8 years respectively) and careful choice of suitable immobilisation matrices is crucial to prevent any environmental contamination. Such an immobilisation material must be able to withstand prolonged heavy ion particle bombardment while maintaining structural integrity. Pyrochlore-type compounds have been proposed as suitable host matrices for this purpose, and great attention has been paid to members of the series $Gd_2(Zr_xTi_{2-x})O_7$ $(0 \leq x \leq 2)$[1,2]. The radiation tolerance of this series increases with increasing zirconium content, and the healing process in the zirconate is expected to be faster than for the titanate as it does not undergo an amorphous transition upon radiation damage and is a fast ion conductor. Devanathan *et al.* have suggested that one of the main factors in the titanate amorphization process is the accumulation of cation Frenkel pairs[3], and we propose a new set of Buckingham potentials, specifically tailored for looking at radiation damage and defect formation in this $Gd_2(Zr_xTi_{2-x})O_7$ series.

THEORY

The pyrochlore structure has Fd3m symmetry and is closely related to that of fluorite. The differences are that the pyrochlore structure has two distinct cation sites and one-eighth of the anion positions are vacant. One-eighth of a unit cell is shown in Figure 1, where the cation and anion sub-lattices are separated for clarity.

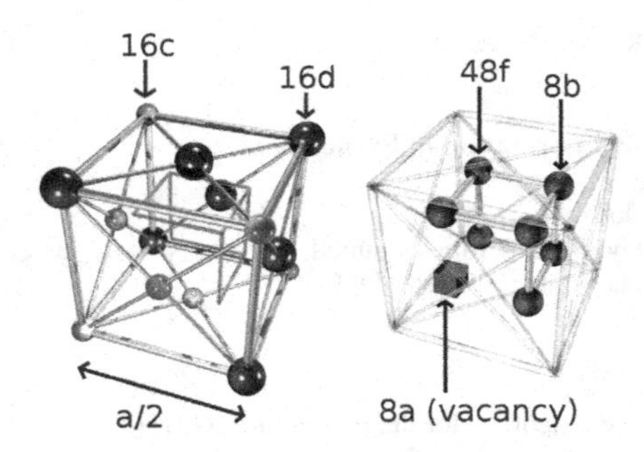

Figure 1. Schematic of one-eighth of the unit cell of a pyrochlore structure. The cation (left) and anion (right) sublattices are separated for clarity. Gd^{3+} ions are in blue, Ti/Zr^{4+} ions are in silver and O^{2-} ions are in red.

New Buckingham potentials are presented (Table I) for the $Gd_2Ti_2O_7$ and $Gd_2Zr_2O_7$ pyrochlores, adapted from our previously published potentials[2] and the Catlow O–O potential[4] by fitting to experimental lattice parameters, ionic distances and elastic constants. The General Utility Lattice Program (GULP)[5] was used for the calculation of structures and energies from the Buckingham potentials. A short range potential cutoff of 12 Å was used and each ion was assigned its formal charge. When calculating the defect energies, the Mott-Littleton approximation[6] was used with radii of 10 Å and 16 Å for regions 1 and 2 respectively.

Table I. Short-range Buckingham potential parameters. The form of the potential is $V_{ij} = A\exp(-r_{ij}/\rho) - C/r_{ij}^6$

Interaction	*A* /eV	*ρ* / Å	*C* / eVÅ^6
O-O	17428.9200	0.1490	27.89
Ti-O	648.6791	0.4065	0.00
Gd-O	1962.7412	0.3250	0.00
Zr-O	806.0606	0.4011	0.00

The activation energies of the oxygen vacancy migrations were calculated using the geometry optimisation and improved tangent nudged elastic band (NEB) method[7] implemented in DL_FIND[8]. The L-BFGS[9] method was used for the optimisation of the NEB path.

The density functional theory calculations were performed using the ABINIT computer code[10], with the generalised gradient approximation (GGA) in the Perdew-Burke-Ernzerhof form[11], the projector augmented wave (PAW) pseudopotential method[12] and a plane-wave cutoff of 15 Ha used for all calculations. Structural optimisation calculations for both the ordered and defective structures used a single **k**-point sampling of the Brillouin zone at the Γ point in a cell consisting of 88 atoms. The influence of different **k**-point sampling and plane-wave cutoff energy was investigated in a series of test calculations and these computational parameters are sufficient to converge the total energy to within 1×10^{-4} Ha. A tolerance of 1×10^{-5} Ha/Bohr is achieved for all structural relaxation calculations, which use the Broyden-Fletcher-Goldfarb-Shanno (BFGS) method for optimisation[13].

DISCUSSION

Lattice parameters, selected bond lengths and calculated elastic constants for non-defective $Gd_2Ti_2O_7$ and $Gd_2Zr_2O_7$ are shown in Tables II and III respectively. Our new pyrochlore potentials give values that are in excellent agreement with experimental values of the lattice parameter and bond lengths, and the elastic constants calculated using the new potentials are similar to those calculated using the distinct oxygen model (DOM) and Bush potentials.

Table II. Calculated and experimental lattice parameters, bond lengths and elastic constants for $Gd_2Ti_2O_7$

	Experiment[14]	**New potentials**	**New *ab initio***	**Old potentials[2]**	**Bush[15]**	**Minervini[16]**	**DOM[17]**	**Wilde[4]**
***a*/Å**	10.185	10.185	10.295	10.191	10.048	10.135	10.228	10.325
d(Gd–Ti)/Å	3.601	3.601	3.640	3.603	3.553	3.583	3.616	3.650
d(Gd–O)/Å	2.459	2.439	2.464	2.471	2.421	2.462	2.471	2.488
d(Ti–O)/Å	1.951	1.965	1.987	1.943	1.928	1.930	1.966	1.981
c_{11}/GPa	—	395.8	-	536.7	443.9	477.5	388.9	360.7
c_{12}/GPa	—	132.7	-	216.3	170.2	191.6	148.5	127.8
c_{44}/GPa	—	86.7	-	205.4	70.1	185.5	71.7	77.2

Table III. Calculated and experimental lattice parameters, bond lengths and elastic constants for $Gd_2Zr_2O_7$

	Experiment[14]	**New potentials**	**New *ab initio***	**Old potentials[2]**	**Minervini[16]**	**DOM[17]**	**Wilde[4]**
***a*/Å**	10.523	10.478	10.700	10.758	10.353	10.232	10.494
d(Gd–Zr)/Å	3.720	3.704	3.783	3.804	3.660	3.618	3.710
d(Gd–O)/Å	2.445	2.482	2.498	2.557	2.365	2.474	2.513
d(Zr–O)/Å	2.093	2.043	2.118	2.091	2.144	1.957	2.025
c_{11}/GPa	—	370.2	-	444.6	131.0	390.7	362.0
c_{12}/GPa	—	113.9	-	171.2	331.0	147.2	123.8
c_{44}/GPa	—	90.9	-	174.0	-461.6	77.5	105.4

The formation energies for a selection of defects are shown in Tables IV and V. All of the energies are reported in values of eV/single point defect, and the Frenkel pairs are all located within the same 1/8th of a unit cell. For the DOM, there are two shell parameterisations when an interstitial oxygen occupies the 8a site, and the use of either the O_{8b} or the O_{48f} model is indicated in parentheses in the table. When looking at cation antisite defects, the 'compact' configuration is modelled, where the cation disorder effectively involves nearest-neighbour switching of a Gd and either Ti or Zr, located around the vacancy. When calculating the Frenkel pair energies in the presence of a cation antisite, the cation disorder is treated as a defect along with the Frenkel pair and all are modelled concurrently. The energy of the cation disorder alone is then subtracted from this total formation energy to obtain the corrected, per-single point defect.

Table IV. Comparison of defect formation energies for $Gd_2Ti_2O_7$ calculated using different potential sets. All values are in eV. Where convergence was not achieved, or where no data were available, entries are marked with a '—'. Where the defective structure relaxed immediately back to the initial structure entries are marked with 'UNS'.

	New potentials	Old potentials[2]	DOM[17]	Bush[15]	Minervini[16]	Wilde[4]
VO_{8b}	21.71	24.88	18.93	18.57	24.27	20.47
VO_{48f}	17.95	21.53	19.07	12.93	21.78	16.97
$O8a_i$	-12.04	-9.46	-12.89 (8b) / -8.71 (48f)	-6.35	-10.99	-11.63
$VO_{8b} + O8a_i$	3.92	6.31	3.10 (8b)	—	5.30	3.29
$VO_{48f} + O8a_i$	UNS	UNS	UNS (48f)	—	UNS	UNS
Split vacancy ($2VO_{48f} + O_i$)	6.00	7.58	6.49 (48f)	-6.63	7.59	5.78
Cation antisite (A/S) pair	2.95	5.52	3.46	-1.18	4.90	3.14
A/S corrected $VO_{8b} + O8a_i$	3.12	5.69	2.65 (8b)	1.11	4.84	1.86
A/S corrected $VO_{48f} + O8a_i$	0.48	-0.04	0.10 (48f)	—	0.60	-0.10

Table V. Comparison of defect formation energies for $Gd_2Zr_2O_7$ calculated using different potential sets. All values are in eV. Where convergence was not achieved, or where no data were available, entries are marked with a '—'. Where the defective structure relaxed immediately back to the initial structure entries are marked with 'UNS'.

	New potentials	Old potentials[2]	DOM[17]	Wilde[4]
VO_{8b}	20.36	21.92	18.93	20.10
VO_{48f}	17.81	21.05	19.32	17.96
$O8a_i$	-14.19	-15.82	-12.98 (8b) / -8.80 (48f)	-13.75
$VO_{8b} + O8a_i$	2.10	1.83	—	—
$VO_{48f} + O8a_i$	0.80	0.11	UNS	2.62
Split vacancy ($2VO_{48f} + O_i$)	5.31	5.98	6.63 (48f)	—
Cation antisite (A/S) pair	1.80	3.13	3.64	—
A/S corrected $VO_{8b} + O8a_i$	0.88	1.70	3.91 (8b)	—
A/S corrected $VO_{48f} + O8a_i$	0.63	-0.25	0.55 (48f)	—

There is good agreement between the different potentials for the formation energies of the single oxygen vacancies and interstitials for both pyrochlores. It should be noted that the DOM treats the two oxygen environments separately with different core and shell charges and different spring constant dependent on whether an oxygen is in the 48f or 8b position. This results in two possible energies whenever an oxygen interstitial is added depending on the parameterisation used. Results for both are given in Tables IV and V. The two values obtained with the DOM potentials for the isolated O_{8a} interstitial shows that the parameterisation dramatically influences the defect energy. A problem arises with the DOM when a new oxygen environment is introduced that has not been parameterised, such as the 8a site. This site can be treated as 8b or 48f when in reality it is quite different from either. The DOM is therefore particularly unsuitable for simulating radiation damage in these materials.

The cation antisite pair formation energy is of interest as the ionic motion and therefore the healing kinetics of these compounds are influenced by an oxygen vacancy-hopping mechanism between 48f sites, and cation disorder is a pre-requisite for the facile formation of appropriate anion vacancies due to a lowering of their formation energy (see Table IV). It has been suggested that the lower cation antisite formation energy in the zirconate does not impede the Frenkel pair recombination as much as in the titanate[3]. In turn, the zirconate is destabilised to a lesser extent than the titanate and does not amorphize. In Table V it can be seen that introducing a single cation antisite defect in $Gd_2Zr_2O_7$ lowers the formation energies of the anion Frenkel pairs supporting this hypothesis of facile Frenkel recombination.

The vacancy-interstitial pair formation energies for the zirconate are lower than in the titanate, and the O_{48f} vacancy-interstitial pair is metastable from the outset (cation disorder is required in the titanate for this). The cation antisite pair formation energy is much lower than in the titanate, in contrast to the DOM where a small increase in formation energy relative to the titanate is seen. This is a consequence of very similar DOM potentials for zirconate and titanate; the sole difference is in the value of the O_{48f} spring constant. The presence of the cation antisite defect again lowers the formation energies of the Frenkel-type defects.

Activation energies for an activated hop of one VO_{48f} to an adjacent such site have been calculated (Table VI). The titanate activation energy of 0.29 eV is in keeping with the values calculated with other potentials (with the exception of Pirzada *et al.*[18]), but remains substantially lower than a value obtained by extrapolation of experimental conductivity values for the series $Gd_2(Zr_xTi_{1-x})_2O_7$ ($x = 0.25$–1.00)[19,20]. It has been suggested[4] that this discrepancy arises because the vacancy formation itself may be more important (an effective rate limiting step) than the migration energetics. The zirconate activation energy of 0.42 eV is broadly in keeping with those obtained with the other potentials. Perfect, ordered pyrochlore structures are designated 0X in Table VI, while one pair of switched Gd and Ti/Zr ions are marked 1X. Looking firstly at the titanate values, it can be seen that a small amount of local cation disorder does not influence the activation energy greatly and the values remain at least 0.5 eV below that found experimentally. With the zirconate the activation energy for the oxygen vacancy migration decreases in the presence of an antisite defect by a significant amount (approximately 0.2 eV) with our new potentials. Williford *et al.*[17] also report a similar significant decrease in the activation energy and, while their absolute values are greater than ours, the percentage decrease in the presence of an antisite defect is similar, at between 40–50%.

Table VI. Calculated values for the activation energy of a $VO_{48f} \rightarrow VO_{48f}$ transition in both $Gd_2Ti_2O_7$ and $Gd_2Zr_2O_7$. All values are in eV. 0X denotes a perfectly ordered structure, and 1X denotes one pair of switched Gd and Ti/Zr ions. Values for the DOM, Wilde and Bush potentials are from calculations performed by Williford *et al.*[17].

	New potentials	**DOM[17]**	**Pirzada[18]**	**Wilde[4]**	**Bush[15]**	**Experiment**
$Gd_2Ti_2O_7$ (0X)	0.29	0.24	1.23	0.33	0.04	0.93[19,20]
$Gd_2Ti_2O_7$ (1X)	0.29	0.47	—	0.26	0.30	—
$Gd_2Zr_2O_7$ (0X)	0.42	1.17	0.58	0.73	—	0.73–0.90[21-23]
$Gd_2Zr_2O_7$ (1X)	0.24	0.63	—	—	—	—

CONCLUSIONS

New Buckingham potentials have been introduced and verified for use with the pyrochlores $Gd_2Ti_2O_7$ and $Gd_2Zr_2O_7$. The activation energies for straightforward $VO_{48f} \rightarrow VO_{48f}$ migrations were calculated for both $Gd_2Ti_2O_7$ and $Gd_2Zr_2O_7$. While the titanate has the lower activation energy for this transition in the nondefective structure, the situation reverses when a small amount of 'compact' cation antisite disorder is introduced. This is consistent with the zirconate possessing a greater radiation tolerance compared to the titanate. A more facile migration of oxygen vacancies in the Zr pyrochlore is likely to lead to faster healing kinetics. These new potentials will be essential for the modelling of radiation damage and associated healing processes in large (~1 million+ atoms) systems as they lack shell components, which are unsuitable for such a task, yet retain a great accuracy in the description of the pyrochlore physical parameters and defect energies.

ACKNOWLEDGMENTS

The authors thank the UK Engineering and Physical Sciences Research Council (EPSRC) for funding under Grant EP/H012990/1.

REFERENCES

1. D.S.D. Gunn *et al.*, *J. Mater. Chem.* **22**, 4675 (2012)
2. J.A. Purton and N.L. Allan, *J. Mater. Chem.* **12**, 2923 (2002)
3. R. Devanathan, W.J. Weber and J.D. Gale, *Energy Environ. Sci.*, **3**, 1551 (2010)
4. P.J. Wilde and C.R.A. Catlow, *Solid State Ionics* **112**, 173 (1998)
5. J.D. Gale, *J. Chem. Soc., Faraday Trans.*, **93**, 629 (1997)
6. N.F. Mott and M.J. Littleton, *Trans. Faraday Soc.* **34**, 485 (1938)
7. G. Henkelman and H. Jónsson, *J. Chem. Phys.* **113**, 9978 (2000)
8. J. Kästner *et al.*, *J. Phys. Chem. A* **113**, 11856 (2009)
9. D.C. Liu and J. Nocedal, *Math. Program.* **45**, 503 (1989)
10. P.-M. Anglade *et al.*, *Computer Phys. Commun.* **180**, 2582 (2009)
11. K. Burke, J.P. Perdew and M. Ernzerhof, *Phys. Rev. Lett.* **77**, 3865 (1996)
12. G. Kresse and J. Joubert, *Phys. Rev. B* **59**, 1758 (1999)
13. C.G. Broyden, *J. Inst. Math. App.* **6**, 76 (1970); R. Fletcher, *Comp. J.* **13**, 317 (1970); D. Goldfarb, *Math. Comp.* **24**, 23 (1970); D.F. Shanno, *Math. Comp.* **24**, 647 (1970)
14. O. Knop, F. Brisse and L. Castelliz, *Can. J. Chem.* **47**, 971 (1969)
15. T.S. Bush *et al.*, *J. Mater. Chem.* **4**, 831 (1994)
16. L. Minervini, R.W. Grimes and K.E. Sickafus, *J. Am. Ceram. Soc.* **83**, 1873 (2000)
17. R.E. Williford *et al.*, *J. Electroceram* **3**, 409 (1999)
18. M. Pirzada *et al.*, *Solid State Ionics* **140**, 201 (2001)
19. H.L. Tuller, *J. Phys. Chem. Solids* **55**, 1393 (1994)
20. S. Kramer, S. Spears and H.L. Tuller, *Solid State Ionics* **72**, 59 (1994)
21. M.P. van Dijk, K.J. de Vries and A.J. Burggraaf, *Solid State Ionics* **9**, 913 (1983)
22. P.K. Moon and H.L. Tuller, *MRS Online Proc. Libr.* **135**, 149 (1989)
23. A.J. Burggraaf, T. van Dijk and M.J. Veerkerk, *Solid State Ionics* **5**, 519 (1981)

Mater. Res. Soc. Symp. Proc. Vol. 1514 © 2013 Materials Research Society
DOI: 10.1557/opl.2013.198

Helium Bubbles in Fe: Equilibrium Configurations and Modification by Radiation

Xiao Gai, Roger Smith and Steven Kenny
Mathematical Sciences Department, Loughborough University, Leicestershire, LE11 3TU, UK

ABSTRACT

We have examined the properties of helium bubbles in Fe using two different Fe-He potentials. The atomic configurations and formation energies of different He-vacancy complexes are determined and their stability in the region of nearby collision cascades is investigated. The results show that the optimal He to Fe vacancy ratio increases from about 1:1 for approximately 5 vacancies up to about 4:1 for 36 vacancies. Collision cascades initiated near the complex show that Fe vacancies produced by the cascades readily become part of the He-vacancy complexes. The energy barrier for an isolated He interstitial to diffuse was found to be 0.06 eV. Thus a possible mechanism for He bubble growth would be the addition of vacancies during a radiation event followed by the subsequent accumulation of mobile He interstitials produced by the corresponding nuclear reaction.

INTRODUCTION

Reduced-activation ferritic/martensitic steels are candidate materials for use in nuclear reactors [1]. The presence of transmutation-created helium plays an important role in the microstructural evolution of these steels under neutron irradiation. Helium has a large effect on cascade damage. Interstitial helium atoms increase the production of Frenkel pairs whilst substitutionals tend to decrease this production [2].

Small helium-vacancy (He_nV_m) clusters may play an important role in the nucleation of He bubbles. However, the atomistic properties of He in metals are difficult to identify experimentally. Thus atomistic simulations such as molecular dynamics (MD) provide useful tools to study the formation and the stability of these clusters. Here, we present the results of a study on the formation of small helium-vacancy clusters in bcc iron and their interaction with nearby collision cascades, which will provide insight into the growth of the bubble. The results were obtained from our in-house MD code, LBOMD.

COMPUTATIONAL METHOD

The formation energies of the helium-vacancy clusters He_nV_m are evaluated using two different Fe-He potentials. The first Fe-He potential is a three-body potential by Stoller *et al.* [3], the other Fe-He potential is a many-body potential by Gao *et al.* [4]. The first potential is combined with the 1997 Ackland *et al.* potential [5] for the Fe-Fe interactions whilst the latter one uses the Ackland and Mendelev potential (AM-potential) for the Fe-Fe interactions [6]. Both of the potentials use the Aziz helium potential for the He-He interactions [7].

The formation energy of an He – vacancy complex is defined as the difference in total energy between a crystal containing a defect and a perfect bcc crystal of the same number of Fe atoms with the corresponding number of helium atoms in the fcc structure. So for n He atoms in a system with m vacancies,

$$E_f = E_b + (n-m)E_{vac} - N_{Fe}E_{Fe} - nE_{He}^{sub}$$

where E_f is the formation energy. E_b is the energy of the lattice containing the bubble, E_{vac} is the formation energy of a single vacancy in bcc-iron, N_{Fe} is the number of Fe atoms in the lattice containing the bubble, E_{Fe} is the cohesive energy of Fe and E_{He}^{sub} is the energy of an helium substitutional atom, defined as follows,

$$E_{He}^{sub} = E_{sub}^{ref} - N_{Fe}^{ref}E_{Fe}.$$

This is the difference between the energy of a reference bcc Fe lattice containing a substitutional He and the number of Fe atoms in the reference lattice multiplied by the cohesive energy of Fe.

In the calculations for the formation energy, the box size L is set to $30a_0$, where a_0 is the lattice parameter. The value of L was chosen so that the energies defined above had converged. For all calculations, periodic boundary conditions and constant volume are used. The clusters were generated with the following procedure. We start by creating the pure bcc Fe lattice, before removing atoms to get an approximately spherical vacancy cluster. Next we generate an He bubble in the fcc structure of approximately the same size and shape as the vacancy cluster and place this into the Fe lattice. The system is then relaxed using a conjugate gradient algorithm. For each He-vacancy ratio several initial random configurations are tested and the one with the lowest formation energy is kept.

To simulate the cascade event, molecular dynamics is used. The system is first thermalised before a cascade event is initiated. We choose a value of 500 K for the system temperature, similar to reactor pressure vessel operating conditions [8-9]. After that a 1 keV cascade is initiated near the bubble shown in figure 1, by imparting 1 keV energy to a primary knock-on atom (PKA). To ensure results are reliable, the chosen cascades must rigorously sample all unique directions within the crystal. With cubic systems, this calculation can be reduced somewhat due to high symmetry resulting in an irreducible volume that is deemed representative of the whole structure but for He bubbles the symmetry is destroyed. Nonetheless we use a similar procedure in the He bubble case. Figure 1 illustrates the range of trajectories chosen for the calculation and the three positions chosen for the PKA.

Since the bcc structure is highly symmetric, the unit can be divided into a few identical tetrahedra, which are illustrated in the right part of figure 1. All the directions can be launched within this tetrahedra due to the symmetry. For a direction <x, y, z>, after first setting $x = 1, y = 0, z = 0$, y and z are increased by a step size of 0.1 keeping $z \leq y$; 66 directions are selected which would give a representative sample of all crystallographic directions in a perfect bcc lattice. Furthermore, for ensuring reasonable statistics, different times of thermalisation are used to provide different initial states of the cascade process. In addition to the cascade calculations we also determined the energy barrier for an isolated He atom to diffuse in a perfect bcc Fe lattice using the dimer method [10].

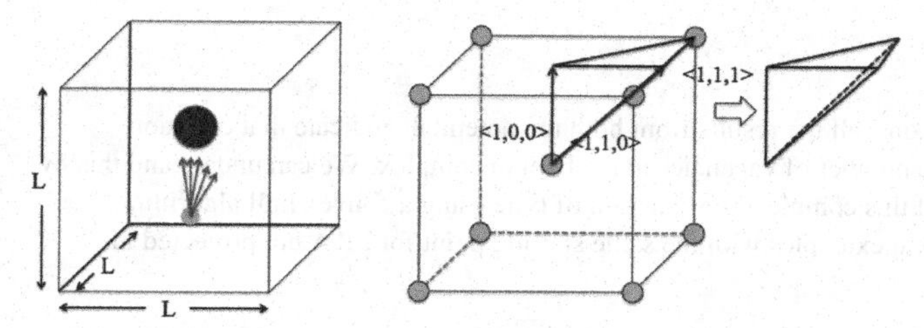

Figure 1 The choice of directions for the collision cascades. The black ball represents the He bubble, whose center is placed offset by $a_0/2$ from the center of the box. The grey sphere represents the PKA, with arrows signifying the directions of the trajectories. The position of PKA here is ($0.3\times L$, $0.5\times L$, $0.5\times L$) but positions at ($0.3\times L$, $0.4\times L$, $0.5\times L$) and ($0.2\times L$, $0.5\times L$, $0.5\times L$) are also chosen. L is the length of the box size, which is equal to $30a_0$.

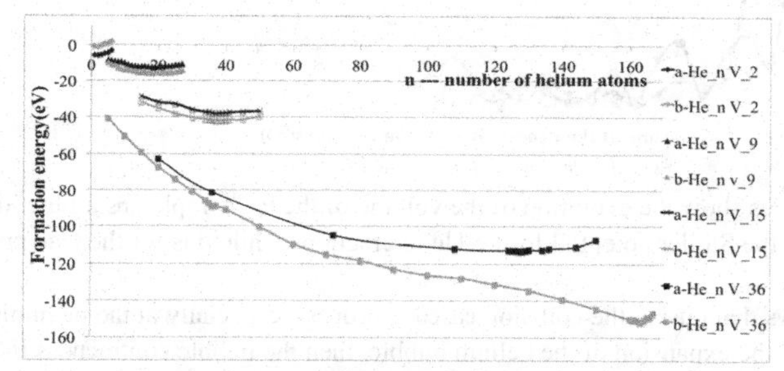

Figure 2. The formation energy as a function of the number of helium atoms for different sizes of He_nV_m clusters. Here, *a* represents the Stoller potential for Fe-He interactions while *b* is for the Gao potential. The optimum ratio corresponds to the minimum of these curves.

RESULTS

FORMATION ENERGY

Figure 2 shows the formation energy calculated from the two different Fe-He potentials. Both empirical potentials reveal the same trends; that is to say there is an optimal He to vacancy ratio for each curve (calculated with a fixed number of vacancies). The optimum ratio increases from around 1:1 for small bubbles up to 4:1 for large bubbles. When comparing the two potentials, it can be seen that the Stoller potential gives a formation energy at least 1.5 eV higher than the Gao potential except for the extremely small clusters (with 2 vacancies) and the optimal ratio for the Gao potential is slightly higher than the one for the Stoller potential. The optimum ratios are similar, except for the larger vacancy clusters where there is some divergence.

CASCADES

Based on the simulations, all the results from both the potentials indicate that collision cascades can increase the number of vacancies in the helium complex. We can understand this by calculating the volume of this complex as a function of time using a convex hull algorithm. Figure 3 shows four typical examples with the same starting point for PKA but projected in different directions.

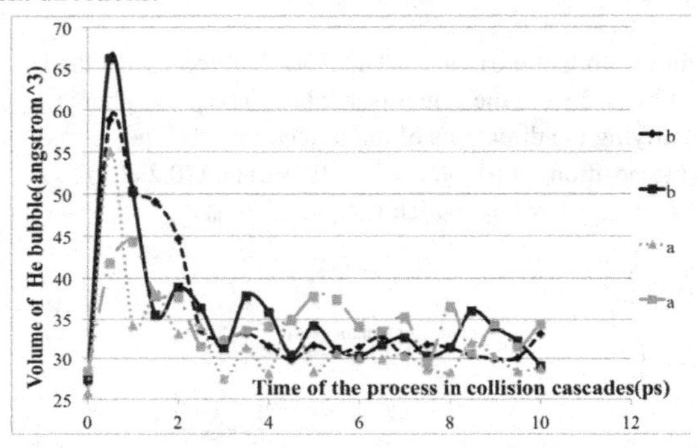

Figure 3. The curves show the evolution of the volume of the He complex as a function of time. Here, *a* represents the Stoller potential for Fe-He interactions while *b* is for the Gao potential.

Figure 3 shows that during the collision cascade process, especially at the beginning, the cascade will cause the expansion of the helium bubble, then the bubble contracts as time goes on and finally its volume becomes stable. Moreover, we find that the volume of the helium bubble is increased compared to the original one when the cascades pass through it, while the volume does not change so much if the cascade is far away from the bubble.

In order to determine the reason for the change in bubble size we examine the increase/decrease in Fe vacancies in the bubble by averaging over all the generated cascades. Figure 4 and figure 5 show examples for three He-vacancy clusters where the number of vacancies is kept constant. As might be expected, at a low helium-to-vacancy ratio, the emission of vacancies is clearly favored. This changes at the optimal helium-to-vacancy ratio, where the bubble absorbs vacancies. When the helium density increases again, this phenomenon becomes more obvious. Both Fe-He potentials demonstrate this conclusion with the Gao potential, showing the effect more strongly than the Stoller potential. Above the optimal ratio, vacancy capture becomes more favored. This is because it is energetically more favorable for vacancies close to the bubble to be absorbed, rather than remain isolated. Figure 6 shows a typical example.

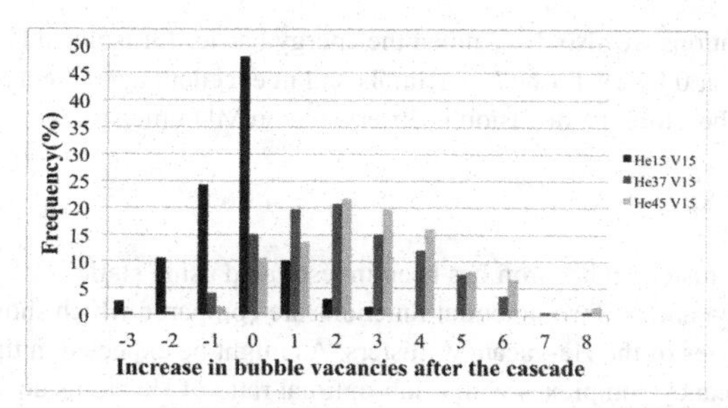

Figure 4. Frequency of capture/loss of vacancies during the collision cascade for a system containing 15 vacancies using the Stoller potential. The three sets of results show three cases of (1) below the ideal He:vacancy ratio, (2) at the optimal ratio and (3) above the ideal ratio.

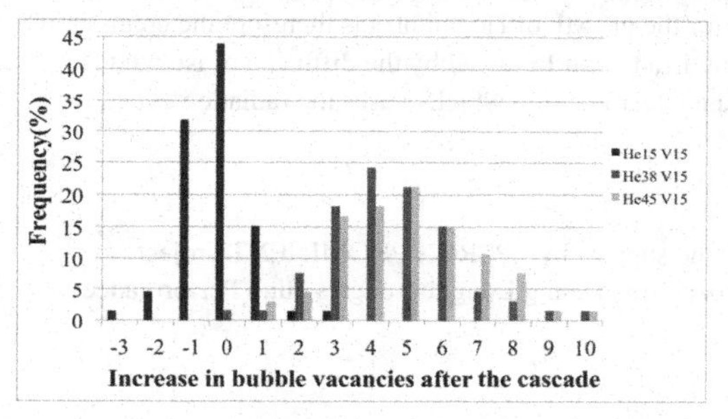

Figure 5. Frequency of capture/loss of vacancies during the collision cascade for a system containing 15 vacancies using the Gao potential. The three sets of results show three cases of (1) below the ideal He:vacancy ratio, (2) at the optimal ratio and (3) above the ideal ratio.

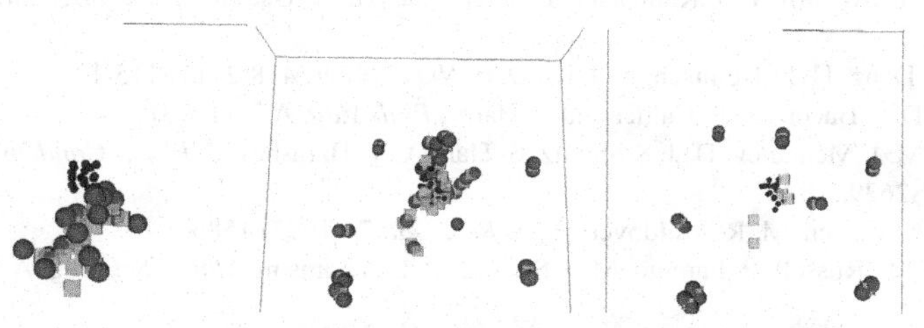

Figure 6. Three frames in the simulation of a collision cascade near a $He_{18}V_9$ complex using the Gao potential. The left image is after 40 fs, just as the cascade reaches the bubble; the center image is after 1200 fs when the cascade has passed into the bubble and the right one is the state after 10 ps, which shows the vacancies (cubes in figure) absorbed by the helium bubble (the small black spheres). The other larger objects are mostly split Fe interstitials.

In addition to the cascade calculations we also determined the energy barrier for isolated He to diffuse was found to be very small at 0.06 eV for both potentials and in excellent agreement with the ab initio calculations [11]. Therefore the diffusion is observable on MD time scales.

CONCLUSIONS

The stability of helium vacancy clusters in bcc iron has been investigated using static atomistic simulation and molecular dynamics. Two potential functions are compared which show similar results for the formation energies of the He-vacancy clusters. As might be expected in the presence of collision cascades He-vacancy complexes with a sub-optimal ratio of He atoms do not attract Fe vacancies but for optimal and super-optimal ratios the cascades enable the complexes to capture vacancies. Furthermore, a very low migration energy (0.06 eV) for diffusion of interstitial He in Fe is obtained, which shows the helium atoms can diffuse over MD time scales. A possible mechanism for the growth of He bubbles is therefore the creation of excess vacancies in the complexes by irradiation followed by the diffusion of isolated He interstitials created in the associated nuclear reaction which causes the radiation event.

ACKNOWLEDGMENTS

This work was carried out with the support by EPSRC's PROMINENT project (EP/I003150/1). Calculations were performed using Loughborough's High Performance Computing Centre.

REFERENCES

1. S. J. Zinkle, *Phys. Plasmas*, 12(2005) 058101.
2. G. Lucas, R. Schäublin, *J. Phys.: Condens. Mater*, 20(2008) 415206.
3. R. E. Stoller, S. I. Golubov, P. J. Kamenski, T. Seletskaia, Yu. N. Osetsky, *Phil. Mag*. 90(2010) 923-934.
4. F. Gao, Huiqiu Deng, H. L. Heinisch, R. J. Kurtz, *J. Nucl. Mater*. 418(2011) 115-120.
5. G. J. Ackland, D. J. Bacon, A. F. Calder and T. Harry, *Phil. Mag*. A 75(1997) p.713.
6. G. J. Ackland, M. I. Mendelev, D. J. Srolovitz, S. Han, A. V. Barashev, *J. Phys.: Condens. Mater*, 16(2004) S2629.
7. R. A. Aziz, A. R. Janzen, M. R. Moldover, *Phys. Rev. Lett.* 74(1995) 1586.
8. M. K. Miller, K. F. Russell, P. Pareige, M. J. Starink, R. C. Thomson, *Mater. Sci. Eng.* A 250(1998) 49.
9. P. Auger, P. Pareige, S. Welzel, J-C. van Duysen, *J. Nucl. Mater*. 280(2000) 331-344.
10. G. Henkelman, H. Jónsson, *J. Chem. Phys.* 111(1999) 7010.
11. C. C. Fu, F. Willaime, *Phys. Rev*. B 72(2005) 064117.

Mater. Res. Soc. Symp. Proc. Vol. 1514 © 2013 Materials Research Society
DOI: 10.1557/opl.2013.517

Atomistic Ordering in Body Centered Cubic Uranium-Zirconium Alloy

Alex P. Moore[1], Ben Beeler[1], Michael Baskes[2,3], Maria Okuniewski[4], and Chaitanya S. Deo[1]
[1] Nuclear and Radiological Engineering Program, George W Woodruff School of Mechanical Engineering, Georgia Institute of Technology, 770 State Street, Atlanta, GA 30332, USA
[2] University of California, San Diego, 9500 Gilman Drive, La Jolla, CA 92093, USA
[3] Los Alamos National Laboratory, PO Box 1663, Los Alamos, NM 87545, USA
[4] Idaho National Laboratory, PO Box 1625, Idaho Falls, ID 83415, USA

ABSTRACT

The metallic binary-alloy fuel Uranium-Zirconium is important for the use of the new generation of advanced fast reactors. Uranium-Zirconium goes through a phase transition at higher temperatures to a (gamma) Body Centered Cubic (BCC) phase. The BCC high temperature phase is particularly important, since the BCC phase corresponds to the temperature range in which the fast reactors will operate. A semi-empirical MEAM (Modified Embedded Atom Method) potential is presented for Uranium-Zirconium. The physical properties of the Uranium-Zirconium binary alloy were reproduced using Molecular Dynamics (MD) simulations and Monte Carlo (MC) simulations with the MEAM potential. This is a large step in making a computationally acceptable fuel performance code.

INTRODUCTION

Metal alloy fuels demonstrate superior performance (over ceramic fuels) in that they behave in a benign manner during core off-normal events, maintain integrity in high burn-up conditions, have low-loss fuel recycling during reprocessing, have a high thermal conductivity, and have a fairly isotropic neutron cross-sections.

Uranium-Zirconium (U-Zr) is a metal alloy that looks to be a promising option as a nuclear fuel. However, for it to be used in a commercial setting, a computational fuel performance code must first be made. Uranium-Zirconium has a Body Centered Cubic (BCC) structure for reactor operating temperatures; therefore, the BCC structure is particularly important to analyze. It is also important to note that the Uranium-Zirconium alloy goes through a δ (C32 Crystal Structure) to γ (BCC) phase transition for 65%-75% Zirconium around 890 Kelvin.

The most recent phase diagram was constructed by H. Okamoto [9] and was made from a compilation of experimental papers on Uranium-Zirconium.

THEORY

MEAM Potential

The MEAM (Modified Embedded Atom Method) inter-atomic potential is a semi-empirical potential proposed by Baskes et al. [11] that has been successfully used to reproduce the physical properties of various metals with different crystal structures. The MEAM potential is useful because it has the ability to replicate physical properties while keeping the computational power and time, which are necessary to complete the simulations, down to an acceptable level. This is a large step in making a fuel performance code and an accurate irradiation damage code.

The MEAM potential for a single element contains 14 adjustable parameters used to obtain the physical properties seen by experiments or *ab-initio* simulations. However, the MEAM potential becomes more complex for binary and tertiary alloys. A binary alloy has 14 adjustable parameters for each element and at least 15 adjustable parameters for the binary alloy interactions.

The MEAM potentials for Uranium and Zirconium [12] [2] have been adjusted for use in the U-Zr alloy. It is important for the MEAM potential of the U-Zr binary alloy to capture some of the physical properties including the crystal structure, thermal expansion, enthalpy of mixing, and phase transitions.

A Modified Embedded-Atom Method (MEAM) potential is presented for the high temperature body-centered cubic (gamma) phase of U. MEAM potentials add an angular component to the older EAM potential to account for directional bonding.

With the MEAM potential, the total energy E of a system of atoms is given by:

$$E = \sum_i \left\{ F_i(\bar{\rho}_i) + \frac{1}{2} \sum_{i \neq j} \phi_{ij}(r_{ij}) \right\}$$

where F_i is the embedding function, ρ_i is the background electron density at site i and $\varphi_{ij}(R_{ij})$ is the pair interaction between atoms i and j at a distance R_{ij}.

The high percent Uranium is not analyzed for this MEAM potential, and some of the more complex phases, which play lesser roles in the physical properties for the temperature range in question, also will not be analyzed.

The parameters for the 2nd Nearest Neighbor MEAM potential are:

Element Parameters for MEAM Potential														
	lat	E_C	α	A	$\beta^{(0)}$	$\beta^{(1)}$	$\beta^{(2)}$	$\beta^{(3)}$	$t^{(0)}$	$t^{(1)}$	$t^{(2)}$	$t^{(3)}$	a_{lat}	ρ^0
U	FCC	5.27	5.1	1.04	6	6.8	7	7	1	2.5	4	3	4.36	1
Zr	BCC	6.2	4.1	0.48	2.8	2	7	1	1	3	2	-7	3.58	1

Table 1: Elemental Modified Embedded Atom Method (MEAM) Potential Parameters

Alloy Parameters for MEAM Potential							
U-Zr	r_e	Δ	α	ialloy	attract(U-U)	attract(Zr-Zr)	attract(U-Zr)
	2.85	0.7	5.5	1	0.1	0	0.04
	r_{cut}	legend	xncut	xmcut	repulse(U-U)	repulse(Zr-Zr)	repulse(U-Zr)
	5.5	0.6	2	6	0.1	0.03	0.04

Table 2: Alloy Modified Embedded Atom Method (MEAM) Potential Parameters

Angular Screening Parameters for MEAM Potential						
	U-U-U	U-Zr-U	Zr-U-U	Zr-Zr-U	Zr-U-Zr	Zr-Zr-Zr
Cmax	1.7	2.8	2.8	2.8	2.8	0.99
Cmin	1.2	0.6	0.8	0.8	0.6	0.7

Table 3: Binary Alloy Modified Embedded Atom Method (MEAM) Potential Screening Parameters

The theory and fitting procedure of a second-nearest neighbor MEAM potential can be found in publications by Byeong-Joo Lee and M. I. Baskes [18] [19]. Most of the elemental parameters are chosen to match experimental, *ab-initio* or Density Functional Theory (DFT) values [11] [17], while the many of the alloy parameters are obtained through trial and error processes [20] [21].

Molecular Dynamics (MD)

The Molecular Dynamics (first principals) simulation consists of a numerical systematic solution to the classical equations of motion. Molecular Dynamics computes the phase-space trajectory, where the atoms are allowed to interact for a period exerting forces on each atom, giving snapshots of the motion of atoms. The forces on each atom are calculated using the MEAM potential developed. Temperature in a Molecular Dynamics simulation corresponds to an average atom velocity.

The Molecular Dynamics Code used, called DYANMO, was run under an Isothermal–isobaric (NPT) ensemble, where the atoms, pressure and temperature are conserved. The initial simulation consists of a periodic random solid solution of 2000 atoms in an un-relaxed perfect BCC lattice.

Monte Carlo (MC)

The Monte Carlo (MC) simulation used is based on the Metropolis Monte Carlo algorithm [13] [14]. The MC method allows for the study of order-disorder and segregation phenomena in the equilibrated system. The MC method is not based on the equations of motion like the MD simulation but the energetics of the states. These types of Monte Carlo simulations are good for evaluating effects that would take a long time to witness during a molecular dynamics simulation.

Order-disorder transitions proceed through substitution between atoms followed by small atomic displacements. These order-disorder transitions are commonly found in metals and alloys. The Monte Carlo (MC) approach is used to drive the atoms toward their equilibrium state at a finite temperature.

The simulation started with the ending positions of each atom after the MD simulation. Then a series of configuration transformations were performed to achieve thermo-dynamically equilibrated state. The Monte Carlo Code uses a canonical (or NVT) ensemble, which means that the number of atoms, volume, and temperature is conserved.

In each MC step, one of the following two configuration changes is attempted with an equal probability:

1. A randomly selected atom is displaced from its original position in a random direction with a distance between 0 and r_{max}.
2. Two randomly selected atoms with different elemental types are exchanged.

Operation (1) accounts for the positional relaxation processes (adjustment of bond lengths and angles), while operation (2) accounts for the compositional relaxation processes (segregation).

$$P_{XY} = \exp\left(\frac{-\Delta E}{k_B T}\right)$$

After each configuration change, we evaluate the energy change between the new and old configurations. If $P_{XY} > 1$ (decrease in energy), the new configuration is accepted.

If $P_{XY} < 1$ (increase in energy), the new configuration is retained with the probability P_{XY}. In the beginning of MC simulations, the potential energy of U–Zr decreases rapidly due to positional and compositional relaxations. However, when the simulations approach equilibrium, there is no significant change in potential energy and the acceptance rate of element exchange operations remains stable around a certain value [14].

Monte Carlo and Molecular Dynamics Methodology

The Monte Carlo (MC) simulation with an NVT ensemble rearranges the atoms to have a lower free energy. This restructuring of the atoms creates a problem since rearranging the atoms should create a volume change. When a Molecular Dynamics simulation is then run with an NPT ensemble, this corrects the volume problem. However, the corrected volume changes how the structure rearranges during an MC simulation. Therefore, an iterative MC-MD simulation is proposed, which should eventually settle to the state that minimizes the free energy through a series of atom switching and thermal motion, if continued. This state may or may not be the lowest energy state since a series of simple atom switching will most likely result in a state which represents a local minimum of the free energy.

A finite number of MC-MD iterations is proposed to approximate the minimal free energy structure.

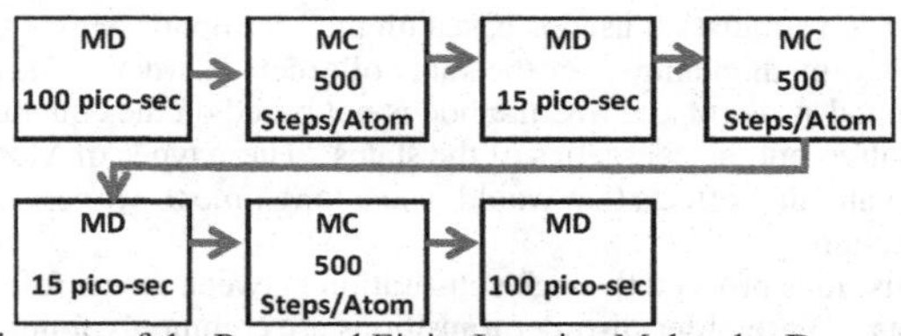

Figure 1: Flow Diagram of the Proposed Finite Iterative Molecular Dynamics (MD) and Monte Carlo (MC) Simulation.

Enthalpy of Mixing (Heat of Formation)

The enthalpy of mixing at zero Kelvin makes the temperature component of the free energy negligible; therefore, the change in the free energy comes from the change in the enthalpy of mixing. While physically, the Body Centered Cubic Phase is unstable at zero Kelvin, to find the temperature effects of the MEAM potential on the enthalpy of mixing, the crystal structure was fixed to remain in the BCC configuration; Dr. Landa in his Monte Carlo and *ab-initio* simulations of Uranium-Zirconium [7] used similar methodology.

For exothermic solutions, $\Delta H_{mix} < 0$, the mixing of the solution results in a free energy decrease. However, the opposite is not necessarily true for endothermic solutions, $\Delta H_{mix} > 0$. The deviations from Vegard's law in the enthalpy of mixing curve can be used to show the expectance of a miscibility gap using the MEAM potential.

$$\Delta H_{mix} = U_{mix} - \sum_i X_i U_i + P_{mix} V_{mix} + \sum_i X_i P_i V_i$$

$$\Delta S_{mix} = -R[X_U \ln(X_U) + X_{Zr} \ln(X_{Zr})]$$

$$\Delta G_{mix} = \Delta H_{mix} - T\Delta S_{mix}$$

Short Range Order Parameter

The Short Range Order (SRO) parameter considers only first Nearest Neighbors.
P_{AA} : is the fraction of the nearest neighbor sites of atom type A that are occupied by A type atoms (averaged over all A atoms).
n_A : is the atomic fraction of A type atoms in the entire system.

$$\sigma = -\frac{P_{AA} - n_A}{1 - n_A}$$

With this definition, σ = 1 for the perfectly ordered lattice, σ = -1 for the phase separated system, and σ = 0 for a random solid solution of equal numbers of A and B atoms. If the number of A and B atoms are unequal, the magnitudes of the extreme values of σ are reduced.

The Short Range Order parameter code assumes a Body Centered Cubic crystal structure, and therefore eight nearest neighbor atoms are also assumed. The order parameter is calculated on a snapshot of the atoms positions at a given time.

RESULTS and DISCUSSION
Enthalpy of Mixing (Heat of Formation)

The MEAM potential used in the simulations results in good agreement with the *ab-initio* and CALPHAD heat of mixing curves from Landa et al. [7]. The positive heat of mixing suggests from Vegard's law the existence of a miscibility gap in the U-Zr phase diagram using the MEAM potential.

The maximum heat of mixing from the MC-MD simulation around $U_{60}Zr_{40}$ is close to the experimental estimated maximum between $U_{70}Zr_{30}$ and $U_{60}Zr_{40}$ [7].

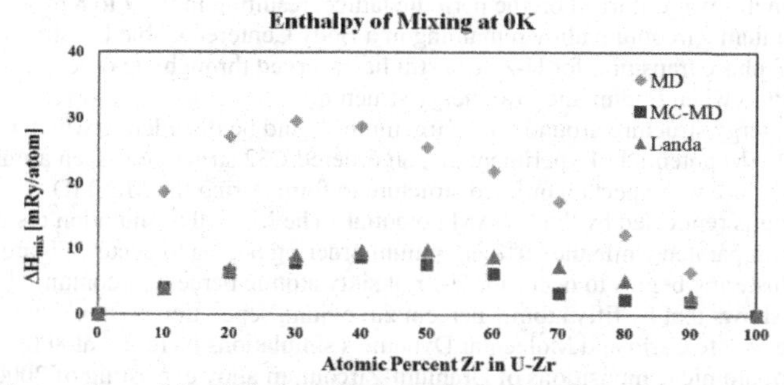

Figure 2: Comparison of the Enthalpy of Mixing (Formation Energy) at Zero Kelvin [7]

The Molecular Dynamics Simulation for a random solid solution lattice parameter (proportional to the thermal expansion) is in good agreement with the experimental lattice parameter value for γ-$U_{30}Zr_{70}$ alloy, within a few percent. The experimental lattice parameter from Landa et al. is 3.589Å for γ-$U_{30}Zr_{70}$ alloy at the γ-δ transition temperature T=925K [7], while the MD simulation lattice parameter was 3.5467Å.

The thermal expansion of the Molecular Dynamics Simulation results in the expected linear curve seen below in Figure 3(a), while the MC-MD thermal expansion has a non-linear nature corresponding to the ordering of the system.

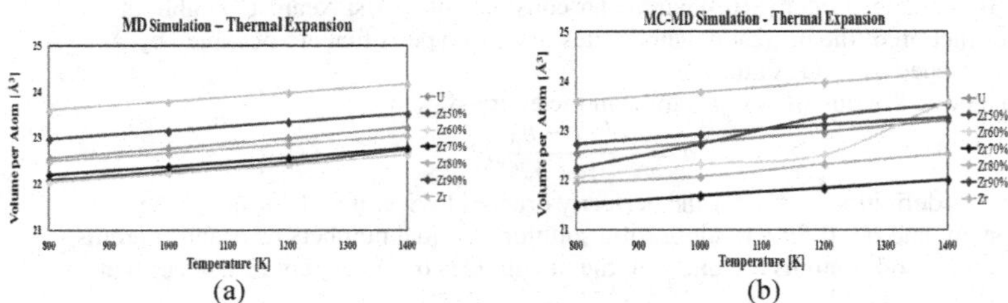

(a) (b)

Figure 3: Thermal Expansion from (a) Molecular Dynamics Simulation of a Random Solid Solution (b) Iterative Molecular Dynamics and Monte Carlo.

Separation (Order/Disorder)

The results of the Molecular Dynamics (MD) simulation against the iterative Monte Carlo and Molecular Dynamics (MC-MD) Simulation relate the difference in properties from a random solid solution versus an ordered or disordered system. Real solutions cannot always be assumed random solutions. Real solutions tend to an atomic arrangement that minimizes the free energy of the mixture.

The order-disorder of the system was seen to play a role in phase stability. In some cases for higher temperature simulations, the order-disorder transition of the atoms improved the stability of the BCC crystal phase where a MD simulation on a random solid solution would have melted.

Forced symmetry was enforced on the periodic lattice, resulting in the γ to δ phase transition of the Uranium-Zirconium alloy remaining in a Body Centered Cubic Crystal structure. The γ to δ phase transition for U-Zr can still be observed through the ordering of the MC-MD simulation when finding the low energy structure.

The lowest energy structure around 50% Zirconium should be disordered, which is replicated by the MEAM potential. Experimentally, an ordered C32 structure is seen around 67% Zirconium, therefore we expect an ordered structure to form during the MC-MD simulations. This too is replicated by the MEAM potential. The MC-MD simulation result at 60% Zirconium is separated while the 70% Zirconium ordering begins to occur. Figure 4 below shows that clustering begins to occur for U-Zr at sixty atomic percent zirconium. In addition, the figure shows that by fifty atomic percent zirconium, separation is seen.

The iterative Monte Carlo and Molecular Dynamics simulations were run at 800 kelvin for an array of atomic compositions of Uranium-Zirconium alloy consisting of 2000 periodic atoms. The final atomic configuration seen in Figure 4 clearly shows the preferential atomic ordering for each composition of Uranium and Zirconium atoms.

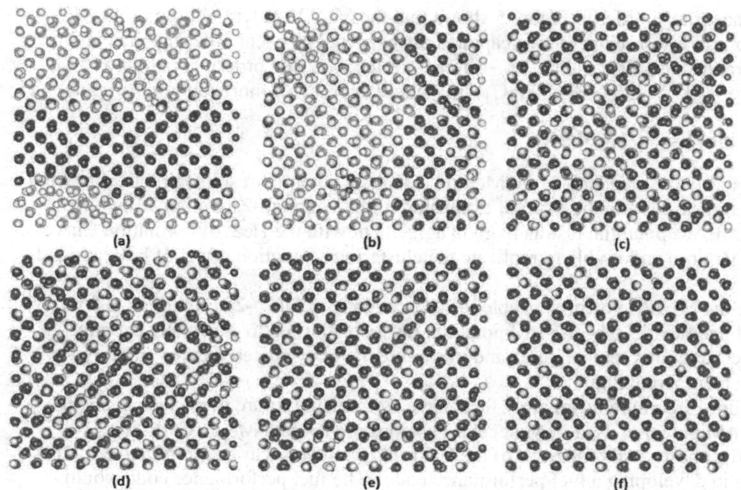

Figure 4: Atomic Arrangements after the Completion of the Iterative Molecular Dynamics and Monte Carlo Simulations at 800K for (a) γ-$U_{60}Zr_{40}$ (b) γ-$U_{50}Zr_{50}$ (c) γ-$U_{40}Zr_{60}$ (d) γ-$U_{30}Zr_{70}$ (e) γ-$U_{20}Zr_{80}$ (f) γ-$U_{10}Zr_{90}$.

The short-range order (SRO) parameter here is a measure of the order versus disorder of the alloy system, once again showing the clustering to segregation effect of U-Zr for lower Zirconium concentrations, and the ordering for higher Zirconium concentrations.

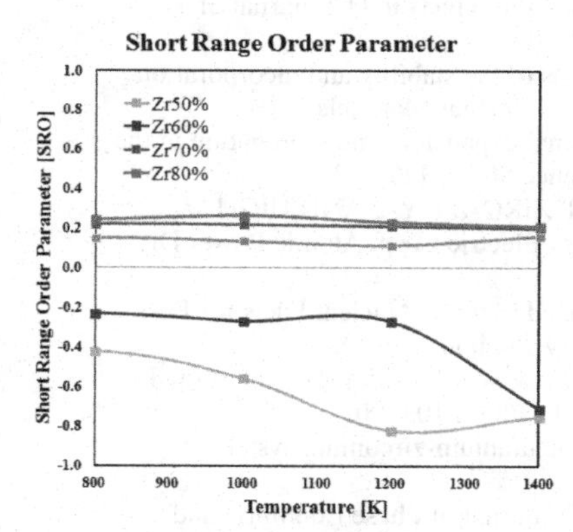

Figure 5: Short Range Order Parameter for Various Atomic Percent Zirconium in the Uranium-Zirconium Alloy.

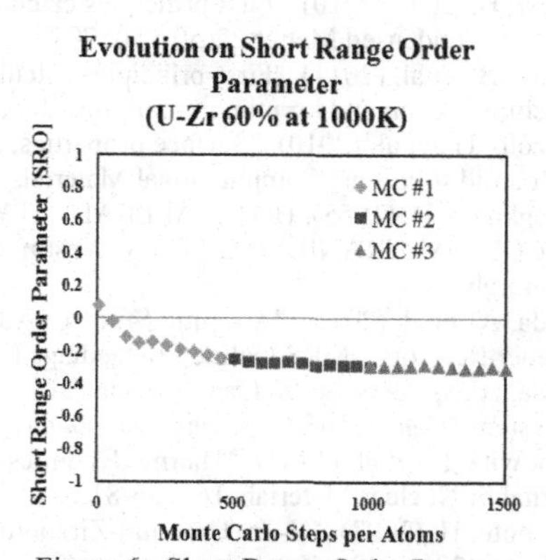

Figure 6: Short Range Order Parameter Evolution throughout the Iterative Molecular Dynamics and Monte Carlo Simulations.

The short-range order in Figure 5 shows that the MEAM potential used in the MC-MD simulations at lower temperatures captures the expected physical ordering. Yet at higher temperatures, $U_{50}Zr_{50}$ and $U_{40}Zr_{60}$ differ from the expected ordering. Around approximately 950K, both $U_{50}Zr_{50}$ and $U_{40}Zr_{60}$ should approach a more random solution.

SUMMARY and CONCLUSIONS

The Modified Embedded Atom Method (MEAM) potential results in replicating both the δ-γ phase transitions and the thermal expansion of the Uranium-Zirconium alloy. In addition, the MEAM potential results in good agreement with the Heat of Formation Curve (Enthalpy of Mixing) and is able to replicate the clustering/separation effect of U-Zr around 800K.

While the MEAM potential is able to capture many of the U-Zr properties, a few scenarios still need work. The MEAM potential must be adjusted to accurately capture ordering effects at higher temperatures and to capture the body centered cubic stability for low atomic percent zirconium.

Uranium-Zirconium metal alloy fuels have a promising future in nuclear science. With the MEAM potential used in Molecular Dynamics (MD) and Monte Carlo (MC) simulations, the phases and properties of the U-Zr alloy can be accurately reproduced. This is a large step in developing a fuel performance code. The fuel performance code would allow the use of U-Zr as a nuclear fuel and allow all of its benefits to be applied to commercial reactors.

REFERENCES

1. Bauer, A. A. (1959). An Evaluation of the Properties and Behavior of Zirconium-Uranium Alloys, Battelle Memorial Inst., Columbus, Ohio.
2. Beeler, B., et al. (2010). "First principles calculations for defects in U." Journal of Physics: Condensed Matter 22(50): 505703.
3. Beeler, B., et al. (2011). "First-principles calculations of the stability and incorporation of helium, xenon and krypton in uranium." Journal of Nuclear Materials.
4. Bozzolo, G., et al. (2010). "Surface properties, thermal expansion, and segregation in the U–Zr solid solution." Computational Materials Science 50(2): 447-453.
5. Droegkamp, R. (1955). HOT MALLEABILITY OF ZIRCALOY-2 AND HIGH ZIRCONIUM-URANIUM ALLOYS, Westinghouse Electric Corp. Atomic Power Div., Pittsburgh.
6. Landa, A., et al. (2012). "Ab Initio Study of Advanced Metallic Nuclear Fuels for Fast Breeder Reactors." MRS Online Proceedings Library 1444(1).
7. Landa, Alex, Per Söderlind, and Patrice EA Turchi. "Density-functional study of the U–Zr system." *Journal of Alloys and Compounds* 478.1 (2009): 103-110.
8. Leibowitz, L., et al. (1989). "Thermodynamics of the uranium-zirconium system." Journal of Nuclear Materials 167: 76-81.
9. Okamoto, H. (2007). "U-Zr (Uranium-Zirconium)." Journal of Phase Equilibria and Diffusion 28(5): 499-500.
10. Rough, F. (1955). An Evaluation of Data on Zirconium-Uranium Alloys, Battelle Memorial Inst., Columbus, Ohio.

11. Baskes, M. I. "Modified embedded-atom potentials for cubic materials and impurities." Physical Review B 46.5 (1992): 2727.
12. Kim, Young-Min, Byeong-Joo Lee, and M. I. Baskes. "Modified embedded-atom method interatomic potentials for Ti and Zr." *Physical Review B* 74.1 (2006): 014101.
13. Metropolis N, Rosenbluth A W, Rosenbluth M N, Teller A H and Teller A E 1953 J. Chem. Phys. 21 1087
14. Wang, Guofeng, et al. "Quantitative prediction of surface segregation in bimetallic Pt–M alloy nanoparticles (M= Ni, Re, Mo)." *Progress in surface science* 79.1 (2005): 28-45.
15. Chevalier, Pierre-Yves, Evelyne Fischer, and Bertrand Cheynet. "Progress in the thermodynamic modelling of the O–U–Zr ternary system." *Calphad* 28.1 (2004): 15-40.
16. Akabori, M., et al. "Stability and structure of the δ phase of the U-Zr alloys."*Journal of nuclear materials* 188 (1992): 249-254.
17. Baskes, M. I., and R. A. Johnson. "Modified embedded atom potentials for HCP metals." Modelling and Simulation in Materials Science and Engineering 2.1 (1999): 147.
18. Lee, Byeong-Joo, and M. I. Baskes. "Second nearest-neighbor modified embedded-atom-method potential." Physical Review B 62.13 (2000): 8564.
19. Lee, Byeong-Joo, et al. "Second nearest-neighbor modified embedded atom method potentials for bcc transition metals." Physical Review B 64.18 (2001): 184102.
20. Jelinek, B., et al. "Modified embedded-atom method interatomic potentials for the Mg-Al alloy system." Physical Review B 75.5 (2007): 054106.
21. Kim, Young-Min, Byeong-Joo Lee, and M. I. Baskes. "Modified embedded-atom method interatomic potentials for Ti and Zr." Physical Review B 74.1 (2006): 014101.

Mater. Res. Soc. Symp. Proc. Vol. 1514 © 2013 Materials Research Society
DOI: 10.1557/opl.2013.122

Cascade Overlap in *hcp* Zirconium: Defect Accumulation and Microstructure Evolution with Radiation using Molecular Dynamics Simulations

Prithwish K. Nandi* and Jacob Eapen**
Department of Nuclear Engineering, North Carolina State University, Raleigh, NC 27695
*pknandi@ncsu.edu, **jacob.eapen@ncsu.edu

ABSTRACT

Molecular dynamics simulations are performed to investigate the defect accumulation and microstructure evolution in *hcp* zirconium (Zr) – a material which is widely used as clad for nuclear fuel. Cascades are generated with a 3 keV primary knock-on atom (PKA) using an embedded atom method (EAM) potential with interactions modified for distances shorter than 0.1 Å. With sequential cascade simulations we show the emergence of stacking faults both in the basal and prism planes, and a Shockley partial dislocation on the basal plane.

INTRODUCTION

Computer simulations of radiation interactions with materials can complement experimental investigations and can aid in developing models for failure of materials especially under irradiation conditions. Zr has been investigated in the past using molecular statics (MS) as well as molecular dynamics (MD) simulations. Willaime and Massobrio [1] performed MD simulations using a tight-binding approach and reported the Burgers mechanism for the temperature dependent *hcp* to *bcc* phase transition. Morris *et al.* [2] calculated the structure and energy of compression twin boundaries in Zr using a combination of first-principles and embedded atom techniques. Serra and Bacon [3] later proposed a new model for {1012} twin growth in *hcp* metals using MD simulations.

Wooding *et al.* [4,5] performed high energy displacement cascade simulations in α-Zr using a Finnis-Sinclair type many–body potential (AWB) that was proposed by Ackland *et. al.* [6]. de Diego and Bacon [7] later probed the defect structures that are produced by near-surface displacement cascades in α-Zr using the same AWB potential. The temperature dependence of creation and clustering of defects generated by displacement cascades in α-Zr was reported by Gao *et al.* [8]. Willaime [9] obtained the formation energies of self-interstitials in *hcp* Zr using density functional theory (DFT) simulations. The mechanisms of interactions between vacancy dislocation loops with the self-interstitial atoms were later reported by Kulikov and Hou [10]. More recently the structure and properties of vacancy and interstitial clusters in α-Zr were investigated by de Diego *et al.* [11] using the AWB potential. Thus it can be noted that the AWB potential has been widely employed in cascade simulations in Zr despite a few well-documented limitations [12]. To overcome some of the drawbacks of the AWB potential [6], Mendelev *et al.* [13] proposed a new EAM potential for α-Zr (M10).

The M10 potential is able to predict the α to β phase transformation in Zr. It also predicts the melting point, liquid structure data and several elastic constants with reasonable accuracy. In this investigation, we have used the M10 potential [13] with an additional hard sphere repulsive interaction at interatomic distances less than 0.1 Å (labeled as M10*) to account for the high compressibility anticipated in displacement cascade simulations [14] at very short times. Ongoing work also considers a spline fit to the Ziegler-Biersack-Littmark (ZBL) potential at short distances that is expected to be more realistic than a hard sphere repulsive potential.

COMPUTATIONAL METHODS

Zr has a hexagonal close-packed (*hcp*) structure [space group: 194 ($P6_3/mmc$)]. For our simulations, we have chosen the orthorhombic representation of the *hcp* lattice with the unit cell spanned by the vectors $\vec{a}_0 = (1,0,0)$, $\vec{b}_0 = \left(0,\sqrt{3},0\right)$ and $\vec{c}_0 = \sqrt{8/3}(0,0,1)$,The orthorhombic unit cell for *hcp* lattice is equivalent to two conventional *hcp* cells with its four basis atoms located at (0,0,0), (1/2,1/2,0), (1/2,5/6,1/2) and (0,1/3,1/2).

A Zr lattice with 256,000 atoms is first relaxed with the conjugate gradient method to its lowest potential energy configuration. It is then equilibrated for 50 ps at constant temperature (300 K) and constant pressure (0 bar) using the LAMMPS [15] MD simulation package. Displacement cascades are then simulated with a 3 keV primary knock-on atom (PKA) introduced close to one of the surfaces of the simulation box. Since it was reported by Wooding et al. [4] that the inclination of the initial PKA direction has no perceptible effect on either the damage state in the collisional phase or the final defect state [4], we have considered the initial PKA momentum perpendicular to the basal plane only. Further investigations will consider other PKA directions. A small set of atoms constituted by three layers of atoms at the system boundary is allowed to absorb the excess thermal energy that crosses into the boundary region using a damped Nose-Hoover thermostat [16]. The boundary atoms also function to attenuate the pressure wave that originates from the radiation knock and prevent it from reentering the simulation box. The atoms in the interior region are allowed to evolve according to the Newton's laws of motion without any constraints. Three different timesteps are used in the cascade simulations – a timestep of 10^{-5} ps is used for the collision phase followed by timesteps of 10^{-4} ps and 10^{-3} ps for the first 3 ps of the annealing phase and the remainder of the annealing phase (25 ps), respectively. For the sequential cascade simulations, the PKA is chosen from the spatial region in the near–vicinity of the first PKA. For each subsequent knock, the PKA momentum and energy are also kept identical to the first knock.

IDENTIFICATION OF DEFECTS

Unambiguous identification of defects in a cascade structure is not straight-forward as there are several criteria available to identify the defected structures. Among the several techniques, Wigner-Seitz cells, the equivalent sphere analysis [17] and common neighbor analysis [18] are widely used for identifying defects in molecular simulations. In this work, we have used the equivalent sphere analysis (ESA) which is based on the Lindemann criterion [17]. In ESA, a defected configuration of the atoms is compared with a reference crystalline state without any defects.

For identifying the vacant sites, each lattice site of the reference configuration is first associated with a sphere. If no atom is found inside the equivalent sphere in the defective configuration, the corresponding lattice site in the reference configuration is tagged as a vacant site. We have used in our analysis a threshold value of 0.32 × lattice constant [7] for the radius of the sphere. A similar procedure is used to estimate the number of displaced atoms. In summary, the ESA method catalogues the vacant sites and the displaced atoms. Note that in this discussion, the term 'displaced atom' points to an atom which is found to be displaced from its equilibrium lattice site.

RESULTS AND DISCUSSION

Equilibrium Properties: We will first report several equilibrium properties with the M10* potential. Table 1 shows the lattice parameters and the cohesive energy of the α-Zr using several potentials including M10*. The lattice parameter at 0 K is determined by performing an energy minimization of the system using the conjugate gradient algorithm. The cohesive energy is defined as the difference between the potential energy of the bound state (crystal) and the isolated atoms which is taken as zero.

Table 1. Equilibrium lattice parameters and cohesive energy per atom for α-Zr with different potentials. M10* refers to M10 potential [14] with an additional hard sphere repulsive potential at distances less than 0.1 Å.

	Experiments	AWB [6]	PM [21]	IKV [22]	M10*
a (Å)	3.232 [19]	3.249	3.232	3.232	3.211
c (Å)	5.147 [19]	5.183	5.149	5.149	5.244
$E_{cohesive}$ (eV)	-6.32 [20]	-6.250	-6.250	-6.250	-6.469

Table 2. Elastic constants, bulk modulus (B), Young modulus (Y), shear modulus (G) and Poisson's ratio (v) for α-Zr with different potentials.

	AWB [6]	PM [21]	IKV [22]	M10*
C_{11} (GPa)	160	146	155	148
C_{12} (GPa)	76	70	67	82
C_{13} (GPa)	70	65	65	63
C_{33} (GPa)	174.7	164.8	173	180
C_{44} (GPa)	36	32	36	48
Y (GPa)				115
Poisson's Ratio (v)				0.3
G (GPa)				44
B (GPa)				96

The stiffness of a material is characterized by its elastic constants which, in the limit of infinitesimal deformation, can be defined as $\sigma_{ij} = \sum_{k,l} C_{ijkl}\varepsilon_{kl}$ where σ_{ij} and ε_{kl} are the elements of the stress and strain tensors, respectively, and C_{ijkl} are the elements of a fourth order elastic constant or stiffness tensor. For a *hcp* system, there are five independent (contracted) elastic constants [23]: C_{11}, C_{11}+C_{12}, C_{13}, C_{33} and C_{44}. In the current study, we have estimated the elastic constants at 0 K with the M10* potential and the results are tabulated in Table 2 (along with a comparison to the predictions from the other potentials). We have also tabulated the values of bulk modulus (*B*), Young's modulus (*Y*), shear modulus (*G*) and Poisson's ratio (v) with the M10* potential in Table 2.

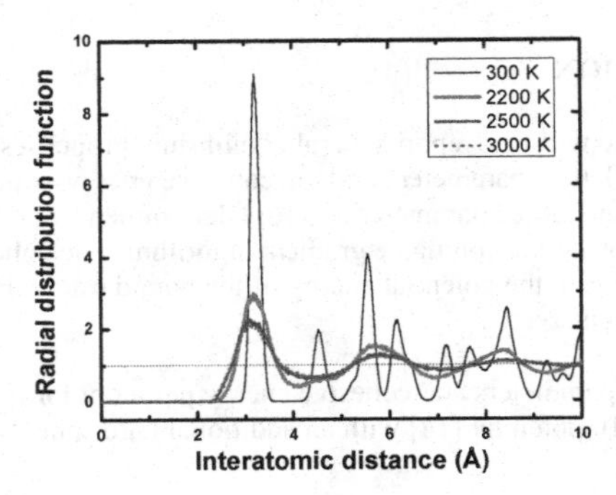

Fig. 1 RDF of Zr at different temperatures.

Radial distribution function: Figure 1 shows the radial distribution function (RDF) of α-Zr at four different temperatures: 300 K, 2200 K, 2500 K and 3000 K with the M10* potential. While the crystalline features are clearly observed at 300 K, the RDF shows a melted disordered state at temperatures above 2200 K. From the changes observed in the RDF, the melting point is estimated to be 2200 K which compares favourably with the experimental value of 2128 K [24].

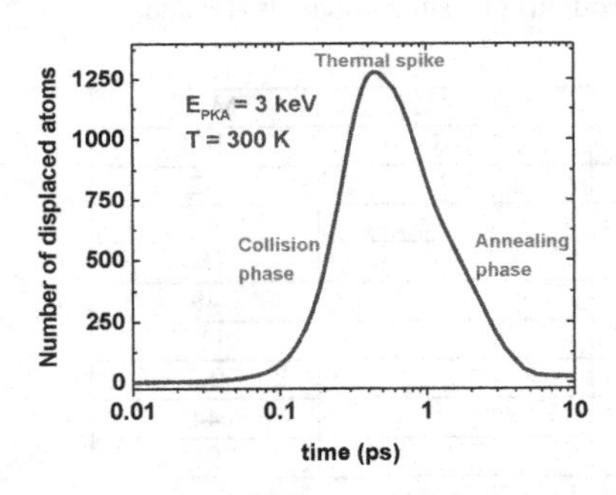

Fig.2 Temporal variation of displaced atoms

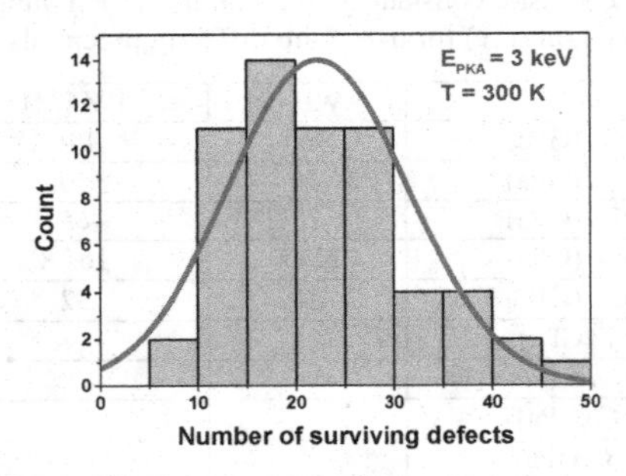

Fig.3 Statistical distribution of displaced atoms

Cascade Simulations: To assess statistically significant defect characteristics, we have employed an isoconfigurational ensemble with copies of the same configuration but with different momenta. The PKA energy (3 keV) and momentum direction (perpendicular to the basal plane) however, remain unchanged in all isoconfigurational copies. The number of displaced atoms (N_D), averaged over 60 independent isoconfigurational copies, is shown in

Figure 2 as a function of time. The number of displaced atoms (N_D) increases rapidly upon the PKA impact (collision phase) and reaches a maximum at the 'thermal spike'. This is followed by quenching or annealing of the system which results in a recombination of defects and a reduction in N_D. At the end of the annealing stage, N_D attains a steady state value. Note that N_D at the end of annealing stage is only a small fraction (2%, approximately) of what is generated at the thermal spike. In Figure 3 we have depicted the statistical distribution of N_D obtained from the isoconfigurational runs. We can observe that there is a wide spread in the number of displaced atoms which survive at the end of the annealing phase. Nevertheless there is evidence for an underlying Gaussian distribution for the number of displaced atoms, a trend which is also observed in cascade simulations in other materials such as copper (preliminary results). More simulations are being conducted on Zr to assess the nature of the distribution.

Defect microstructure from cascade overlap: A main objective of the present study is to investigate the evolution of extended defects such as stacking faults and dislocations in Zr under irradiation with cascade overlap. Therefore, we irradiate the same sample several times sequentially after allowing the simulation to achieve a constant number of displaced atoms at the end of the annealing stage for each knock. Typically, the number of displaced atoms becomes constant in approximately 10 *ps*. In addition, the system is allowed to equilibrate for 15 *ps* before the initiation of each subsequent radiation knock. Note that the timescale for these MD simulations are generally smaller than what is accessible in most experiments. The resulting microstructure is analyzed using an in-house code as well as with DXA [25] for identifying point and extended defects.

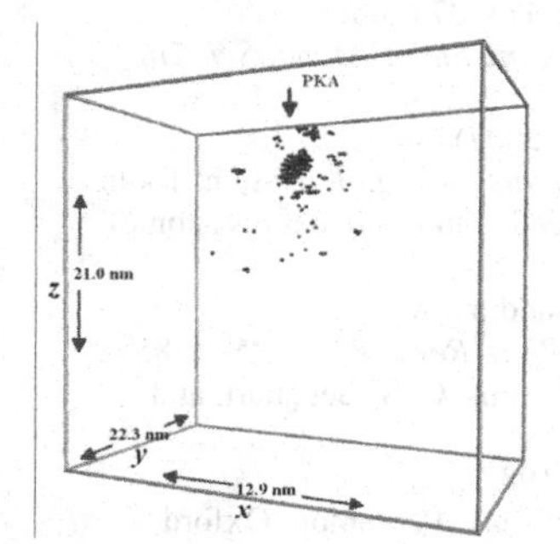

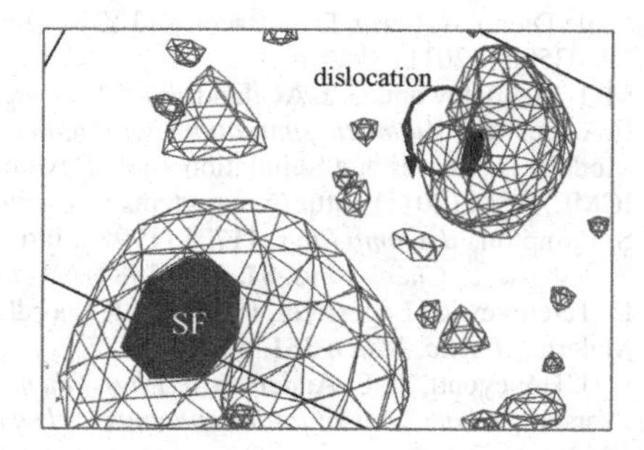

Fig.4 (Left) Defect microstructure from cascade overlap. Green spheres represent interstitials and red spheres represent vacancies. (Right) Stacking faults (SF) (both in the basal and prism planes), and a Shockley partial dislocation with a (1/6)<112> Burgers vector on the basal plane. The small structures shown as blue colored mesh (right) represent an interstitial or vacancy surface.

Figure 4 shows the defects generated from an overlap of six sequential cascades. By and large we have observed that defects are primarily comprised of single interstitials, vacancies and split vacancies following a single knock. With sequential knocks, we see the emergence of stacking faults (SF) both in the basal and prism planes. We have also observed a Shockley partial dislocation having a (1/6)<112> Burgers vector on the basal plane.

CONCLUSIONS

We have performed MD simulations to investigate the microstructure evolution in *hcp* zirconium using an Embedded Atom Method (EAM) potential with a hard sphere repulsion for distances shorter than 0.1 Å. Our studies show that point defects, as well as extended defects such as stacking faults and dislocation loops, can be generated by sequential cascade interactions. Of particular interest is the emergence of stacking faults both in the basal and prism planes, and a Shockley partial dislocation on the basal plane.

REFERENCES

1. F. Willaime and C. Massobrio, *Phys. Rev. Lett.* **63,** 2244 (1989).
2. J. R. Morris, Y. Y. Ye, K. M. Ho, C. T. Chan and M. H. Yoo, *Phil. Mag. A* **72,** 751 (1995).
3. A. Serra and D. J. Bacon, *Phil. Mag. A* **73,** 333 (1996).
4. S. J. Wooding and D. J. Bacon, *Phil. Mag. A* **76**, 1033 (1997).
5. S. J. Wooding, L. M. Howe, F. Gao, A. F. Calder and D. J. Bacon, *J. Nucl. Mater.* **254,** 191 (1998).
6. G. J. Ackland, S. J. Wooding and D. J. Bacon, *Phil. Mag. A* **71,** 553 (1995).
7. N. de Diego and D. J. Bacon, *Phil. Mag. A* **80,** 1393 (2000).
8. F. Gao, D. J. Bacon, L. M. Howe and C. B. So, *J. Nucl. Mater.* **294,** 288 (2001).
9. F. Willaime, *J. Nucl. Mater.* **323,** 205 (2003).
10. D. Kulikov and M. Hou, *J. Nucl. Mater.* **342,** 131 (2005).
11. N. de Diego, Y. N. Osetsky and D. J. Bacon, *J. Nucl. Mater.* **374,** 87 (2008).
12. N. de Diego, A. Serra, D. J. Bacon and Y. N. Osetsky, *Modelling Simul. Mater. Sci. Eng.* **19,** 035003 (2011).
13. M. I. Mendelev and G. J. Ackland, *Phil. Mag. Lett.* **87,** 349 (2007).
14. C.A. Becker, "*Atomistic simulations for engineering: Potentials and challenges*" in Tools, Models, Databases and Simulation Tools Developed and Needed to Realize the Vision of ICME, ASM (2011). http://www.ctcms.nist.gov/potentials
15. S. Plimpton, *J. Comp Phys.* **117**, 1 (1995). http://lammps.sandia.gov
16. S. A. Nose, *J. Chem. Phys.* **81,** 511 (1984).; W. G. Hoover, *Phys. Rev. A* **31**, 1695 (1985).
17. D. Terentyev, C. Lagerstedt, P. Olsson, K. Nordlund, J. Wallenius, C. S. Becquart, and Malerba, *J. Nuc. Mater.* **351**, 65 (2006).
18. D. J. Honeycutt, H. C. Andersen, *J. Phys. Chem.* **91**, 4950 (1987).
19. Pearson., *A handbook of lattice spacings and structures of metals* (Pergamon, Oxford, 1967).
20. C. Kittel, , *Introduction of Solid State Physics*, (NY:Wiley, 1986).
21. R. C. Pasianot and A. M. Monti, *J. Nucl. Mater* **264**, 198 (1999).
22. M. Igarashi, M. Khantha and V. Vitek, *Phil. Mag. B* **63**, 603 (1991).
23. S. Pronk and D. Frankel, *Phys. Rev. Lett.* **90**, 255501 (2003).
24. http://www.webelements.com/zirconium/
25. A. Stukwoski and K. Albe, *Modelling Simul. Mater. Sci. Eng.* **18,** 085001 (2010).

Mater. Res. Soc. Symp. Proc. Vol. 1514 © 2013 Materials Research Society
DOI: 10.1557/opl.2013.61

Role of CSL Boundaries on Displacement Cascades in β-SiC

Prithwish K. Nandi*, V. Ajay Annamareddy and Jacob Eapen**
Department of Nuclear Engineering, North Carolina State University, Raleigh, NC 27695
*pknandi@ncsu.edu, **jacob.eapen@ncsu.edu

ABSTRACT

Molecular dynamics (MD) simulations are carried out to understand the mechanisms of damage production and recovery near grain boundaries in β-SiC under neutron irradiation. Our investigations show that the damage generated by radiation is reduced by the presence of a ∑9{122}[110] tilt grain boundary. Directional displacements which are averaged over an isoconfigurational ensemble are used to characterize the statistical nature of atomic mobility near the grain boundary.

INTRODUCTION

Silicon carbide is considered as an advanced clad/structural material for current light water reactors and the next generation nuclear reactors [1-5]. One of the main concerns of using SiC in reactors stems from the large volumetric swelling and amorphization under prolonged irradiation [6]. Therefore, there is an increased interest in exploring ways to augment the radiation tolerance behavior in SiC through experiments and computer simulations.

Radiation tolerance is generally associated with the ability of a material to withstand the deleterious effects of radiation induced phenomena such as amorphization, point–defect clustering, vacancy induced cavitation, hardening, loss of ductility, swelling, dimensional instability and precipitation of new phases. Following a collision cascade, the freely migrating point defects escape the cascade zone and contribute to the macroscale defect structures. The accumulation of defects is regarded as a competition between defect formation by cascades, thermal fluctuations, interaction between the defects, and the transport or migration of defects. Conversely, the removal of radiation–induced point defects, in principle, can be instrumental in improving the response of a material to intense radiation exposure.

Several investigations show that grain boundaries are instrumental in reducing the point defects which are formed during the cascade simulations [7-9]. Thus materials with a large volume fraction of grain boundaries are often considered to be good candidates for resisting radiation induced damage. Recently, Zhang *et al.* [6] have shown that engineered nanocrystalline SiC with a high density of stacking faults promotes efficient recombination of point defects which results in enhanced resistance to amorphization. A computational study by Swaminathan et al. [10] with various tilt grain boundaries in β-SiC, on the other hand, has shown that the damage in the in-grain region is not affected by the grain boundary type and the magnitude of damage is comparable to the damage in single crystals under normalized conditions.

In the current study, we have investigated the interaction of displacement cascades generated by a primary knock-on atom (PKA) with a ∑9{122}[110] tilt grain boundary in β-SiC using molecular dynamics (MD). A key thrust is on the investigation of the statistical nature of atom mobility near the interface using an isoconfigurational ensemble.

GRAIN BOUNDARY (GB) MODEL

Experiments using high resolution electron microscopy (HRTEM) show that SiC admits a variety of stable grain boundaries [11-15]. One of the interesting features of SiC grain boundary is the presence of 'wrong' bonds – bonds between Si atoms and bonds between C atoms that change the stoichiometry of the interface [11].The geometrical concept of a grain boundary can be described as follows [16]: Consider two identical crystals – *A* and *B*. If crystal B is rotated around a common axis *P* by an angle □ such that for the superposition of the two crystals, the interface is defined by its normal $\hat{n}$ and its location defined with respect to crystal *A*. If the normal to the interface $\hat{n}$ is parallel to *P*, the interface is named as a tilt boundary and if *P* and $\hat{n}$ are perpendicular to each other, it is named as a twist boundary. If the interface planes for the two grains in a tilt boundary are crystallographically identical, the interface is called a symmetric tilt boundary.

In principle, there are many possibilities of orientations between two grains. Among the possible orientations, a few of them are regarded as special for which the GB boundary is described as a coincident site lattice (CSL) boundary [16]. A CSL can be described as a superposition of the two crystals (crystal A and crystal B) rotated around the axis P with a common origin, such that the intersection consists of coincident lattice points – the lattice points which are common to both of the lattices. Generally, a CSL is designated as $\sum\{hkl\}[hkl]$ where $\sum$ is defined by the ratio of the unit-cell volume of the CSL to that of the original lattice, $\{hkl\}$ is the mirror plane and the $[hkl]$ is the rotation axis. CSL boundaries are extensively studied as they are frequently found in experiments and they can be handled well with computational techniques. In the present study we have used a *p*-type $\sum 9\{122\}[110]$ tilt grain boundary which is shown in Figure 1.

Figure 1. $\sum 9\{122\}[110]$ tilt grain boundary in β-SiC

MOLECULAR DYNAMICS (MD) SIMULATIONS

MD simulations are performed using LAMMPS [17] package developed at the Sandia National Laboratories. The interatomic interactions are described by a Tersoff potential optimized by Devanathan *et al.* [18]. In order to obtain the most stable bicrystal configuration, the energy of the atoms is minimized using a conjugate gradient (CG) algorithm. The lowest energy configuration, thus obtained, is then equilibrated (for 50 *ps*) at the desired temperature and zero pressure using a NPT ensemble.

For radiation cascade simulations, a primary knock-on atom (PKA) is given additional momentum perpendicular to the grain boundary as indicated in Figure 2. Fixed boundary conditions are applied for the boundaries parallel to the GB plane, while periodic boundary conditions are used in the other directions. In each grain, three distinct regions are defined as shown in Figure 2. In Region 3, the relative positions of the atoms are kept constant throughout the simulation while in the other two regions, the atoms are free to move. In Region 2, the atoms are coupled to a Nose-Hoover thermostat to absorb the excess thermal energy that is generated by the PKA while the atoms in Region 1 move according to the Newton's laws of motion without any constraints. In this investigation, the PKA (Si atom) is introduced with a 5 keV initial energy in a simulation box consisting approximately 400,000 atoms.

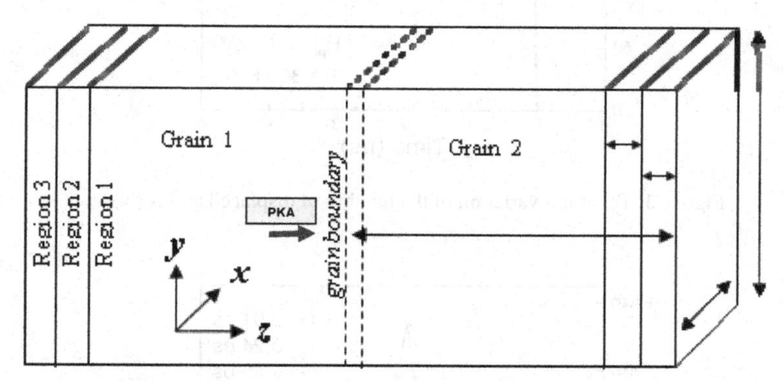

Figure 2. Schematic of MD simulation cell for cascade simulations

ISOCONFIGURAL ENSEMBLE

One of the objectives of the current work is to assess the correlated dynamics near the interface following a radiation impact. In particular, we pose the following question: – what is the average displacement of each atom in a particular direction after the impact? To assess a statistically meaningful displacement metric, we employ an isoconfigurational ensemble with copies of the same configuration but with different momenta. The knock energy and direction however, remain unchanged in all the copies. Then the displacement of each atom is averaged over all the copies of the isoconfigurational ensemble.

RESULTS AND DISCUSSIONS

The number of displaced atoms generated in a displacement cascade for a single crystal is contrasted with that for a bicrystal in Figure 3. The data is averaged over ten independent cascades in the isoconfigurational ensemble. It is interesting to note that while the number of displaced atoms at the thermal spike in both crystals is approximately the same, the number of displaced atoms which survive at the end of the annealing phase is less for the bi-crystal. This indicates that the bi-crystal promotes more annihilation of the displaced atoms during the annealing phase.

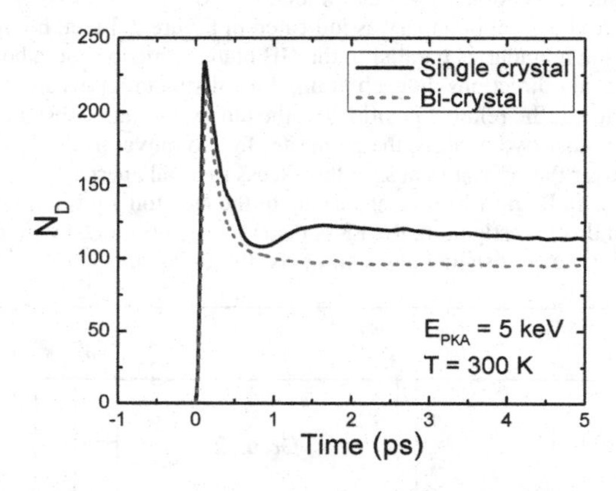

Figure 3. Temporal variation of the number of displaced atoms (N_D)

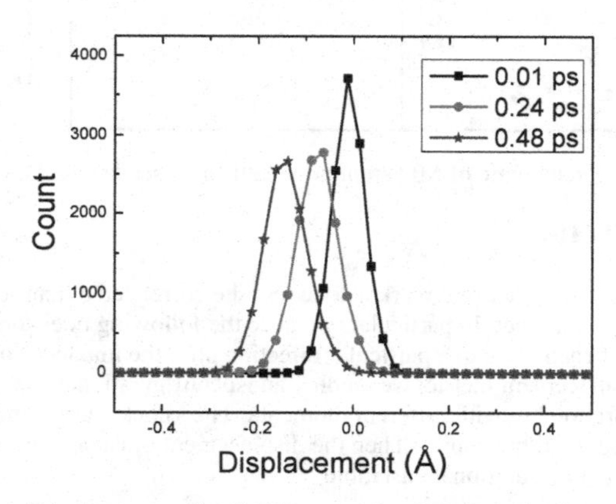

Figure 4. Displacement distribution near the GB along the knock (*z*) direction

Figure 3 does not give an indication on the location of the recombinations in the bi-crystal. To ascertain whether the GB interface promotes recombination we have evaluated the atom migration near the GB using the isoconfigurational ensemble. We first define a displacement metric which is defined as

$$\left\langle \Delta s_i(t)_{ic} = \frac{1}{M}\sum_{k=1}^{M}\Delta s(i,k,t) \right\rangle \quad (1)$$

where s denotes directional (signed) displacement in x, y or z direction, and M is the number of copies of the isoconfigurational ensemble. Note that the metric in Equation (1) is a measure of average displacement of a particular atom (i).

The displacement distribution in the knock direction (averaged over 40 isoconfigurational copies) for the atoms near the GB interface is shown in Figure 4. Interestingly we observe an average displacement in the negative z direction as time progresses. Since the knock direction is in the +z direction, we can conclude that several displaced atoms near the GB start moving towards the GB at short timescales. This reverse migration stems from the correlated dynamics at the interface and results in more annihilation of the displaced atoms at the GB interface. Note that the displacement distributions shown in Figure 4 are calculated over the atoms located inside a hypothetical cubic box placed at the GB (around 20,000 atoms). This ensures that the average negative displacement is not due to any contribution from the relatively less displaced atoms, away from the interface. Two snapshots corroborating the GB recombination are shown in Figure 5.

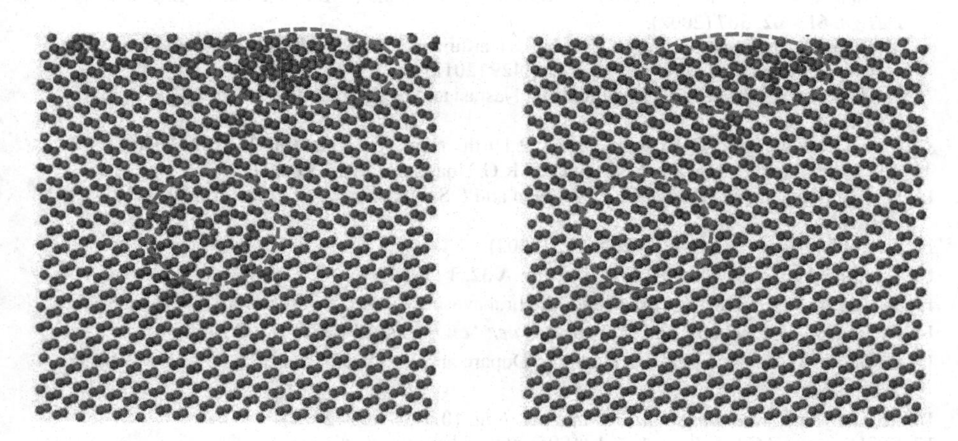

Figure 5. Snapshots of the interface configuration during (left) thermal spike, and (right) annealing phase

CONCLUSIONS

Using molecular dynamics (MD) simulations we have investigated the interaction of grain boundaries with displacement cascades generated by a primary knock-on atom in β-SiC. The interatomic interactions are described by a Tersoff potential which has been optimized by Devanathan et al. [18]. Our investigations show that the damage generated by radiation is reduced by the presence of a $\sum 9\{122\}[110]$ tilt grain boundary. Directional displacements which are averaged over an isoconfigurational ensemble are used to characterize the statistical nature of atomic mobility near the grain boundary.

ACKNOWLEDGEMENT

We acknowledge interesting discussions with W. Weber at the University of Tennessee, Knoxville.

REFERENCES

1. G. C. Rybicki, *J. Appl. Phys.* **78,** 2996 (1995).
2. M. A. Capano, R. J. Trew, *Mater. Res. Soc. Bull.* **22,** 19 (1997).
3. W. J. Choyke, G. Pensl, *Mater. Res. Soc. Bull.* **22**, 25 (1997).
4. A. R. Raffray, R. Jones, G. Aiello, M. Billone, L. Giancarli, H. Golfier, A. Hasegawa, Y. Katoh, A. Kohyama, S. Nishio, B. Riccardi and M. S. Tillack, *Fusion Eng. and Design.* **55,** 55 (2001).
5. L. Giancarli, H. Golfier, S. Nishio, R. Raffray, C. Wong and R. Yamada, *Fusion Eng. and Design.* **61 – 62**, 307 (2002).
6. Y. Zhang, M. Ishimaru, T. Varga, T. Oda, C. Hardiman, H. Xue, Y. Katoh, S. Shannon and W. J. Weber, *Phys. Chem. Chem. Phys.* **14**, 13429 (2012).
7. X-M. Bai, A. F. Voter, R. G. Hoagland, M. Nastasi and B. P. Uberuaga, *Science* **327**, 1631 (2010).
8. M. J. Demkowicz, R. G. Hoagland and J. P. Hirth, *Phys. Rev. Lett.* **100**, 136102 (2008).
9. A. Misra, M.J. Demkowicz, X. Zhang and R.G. Hoagland, *JOM* **59**, 62 (2007).
10. N. Swaminathan, M. wojdyr, D. D. Morgan and I. Szlufarska, *J. App. Phys.* **111**, 054918 (2012).
11. C. Kohler, *Phys. Stat. Sol. (b)* **234**, 522 (2002)
12. K.Hiraga, *Sci. Rep. Res. Inst. Tohoku Univ.* **A 32**, 1 (1984).
13. S. Hagege, D. Shindo, K. Hiraga and M. Hirabayashi, *J. Phys. IV Colloq.* **51**, C1-167 (1990).
14. K. Tanaka, M. Kohyama and M. Iwasa, *Mater. Sci. Forum* **294-296**, 187 (1999).
15. C. Godon, C. Ragaru, O.B. M. Hardouin Duparc and M. Lancin, *Mater. Sci. Forum* **294-296**, 277 (1999).
16. M. Kohyama, *Modelling Simul. Mater. Sci. Eng.* **10**, R31-R59 (2002).
17. S. Plimpton, *J Comp Phys*, **117**, 1 (1995). (http://lammps.sandia.gov/)
18. R. Devanathan, T. Diaz. De la Rubia and W. J. Weber, *J. Nucl. Mater.* **253**, 47 (1998).

Mater. Res. Soc. Symp. Proc. Vol. 1514 © 2013 Materials Research Society
DOI: 10.1557/opl.2013.449

An attempt to handle the nanopatterning of materials created under ion beam mixing

D. Simeone[1], G. Baldinozzi[2], D. Gosset[2], G. Demange[2], Y. Zhang[3], L. Luneville[4]
[1]DEN/DANS/DMN/SRMA/LA2M/LRC-CARMEN, CEA Saclay, 91191 Gif-sur-Yvette, France
[2]CNRS-SPMS/UMR 8580/ LRC CARMEN Ecole Centrale Paris, 92295 Châtenay- Malabry
[3]Materials Science & Technology Division, Oak Ridge National Laboratory, Oak Ridge, Tennessee 37831, USA
[4]DEN/DANS/DM2S/SERMA/LLPR/LRC-CARMEN, CEA Saclay, 91191 Gif-sur-Yvette, France

ABSTRACT

In the past fifty years, experimental works based on TEM or grazing incidence X ray diffraction have clearly shown that alloys and ceramics exhibit a nano pattering under irradiation [1,2,3]. Many works were devoted to study the nano patterning induced by ion beam mixing in solids [17,18,19]. Understanding the nano patterning will provide scientific bases to tailor materials with well-defined microstructures at the nanometric scale. The slowing down of impinging particles in solids leads to a complex distribution of subcascades. Each subcascade will give rise to an athermal diffusion of atoms in the medium. In this work, we focused on this point. Based on the well-known Cahn Hilliard Cook (CHC) equation, we analytically calculate the structure factor describing the nano patterning within the mean field approximation. It has shown that this analytical structure factor mimics the structure factor extracted from direct numerical simulations of the time dependent CHC equation. It appears that this structure factor exhibits a universal feature under irradiation.

INTRODUCTION

It is now well established that materials under irradiation exhibit unusual patterns [1,2,3,4]. Ion solid interaction is of significant interest to both academic and industrial researchers [1]. Ion implantation revolutionized the microelectronic industry offering a control over the number and depth of doping atoms in semiconductor materials [1]. Nowadays, the development of high current and high voltage implanters allows to tailor new compounds with new properties at the nanometric scale [5,6]. These unusual properties result from a steady state pattern formation induced by the slowing down of impinging particles under irradiation. From its ability to modify the local order over few nanometers, ion beam mixing appears to be a promising tool. However, elementary mechanisms responsible for this patterning are far to be clearly understood. Understanding the various mechanisms giving rise to both equilibrium and non equilibrium pattern formation in complex systems is a problem of long standing interest [7].

Two main reasons explain this lack of understanding. The slowing down of incident particles (ions, neutrons) leading to the creation of highly damaged area, termed thermal spikes or subcascades, is a stochastic process difficult to handle [8,9]. On the other hand, it remains difficult to handle the effect of a thermal spike on the microstructure of materials. From the seminal work of Martin and Bellon [10,11], it seems now well established that the effect of a thermal spike on the microstructure can be simulated by an athermal particle exchange. Such a description can also be derived from the seminal work of Sigmund on the slowing down of particles in matter [12].

In this paper, we will focus on the second point and we will discuss the competition between two mechanisms: on one hand, an athermal displacement of atoms induced by a subcascade and on the other hand, the usual thermal driven mechanism trying to bring the solid to the thermo dynamical equilibrium. The steady state microstructure results from the balance between these two mechanisms. The first mechanism leads to destroy the long range order whereas the second tends to restore the long range order.

In the first part of the paper, we describe the effect of an athermal diffusion of atoms in a subcascade within the Landau framework of the Time Dependent Ginzburg Landau (TDGL) equation, extensively used to study the evolution of materials upon irradiation. The main interest of this work is to show critical parameters describing the subcascades able to generate a steady state nano patterning. Within this framework, we calculate the structure factor using a mean field approximation. In the second part of this work, we perform direct numerical simulations to compute this structure factor. The comparison between two structure factors allows assessing the different assumptions. In the last part of the text, we discuss the shape of the structure factor versus the irradiation parameters W and R.

MODELLING A SUBCASVADE WITHIN THE TDGL EQUATION FRAMEWORK

Even if at the microscopic level, the evolution of a material under irradiation can be described by an Ising model with a Glauber like spin flip kinetic, it remains possible to describe at a coarse grained level the microstructure of a material in terms of order parameters. Such a description was used to explain for instance the appearance of tetragonal zirconia nano crystals as well as the fragmentation of spinels under irradiation [2,4,12]. Such description also allows pointing out the amorphisation processes in glasses [13]. The TDGL was extensively used to describe the mechanism for phase separation in binary alloys. Cahn and Hilliard first introduced the conservative order parameter $\eta(r,t)=c_A(r,t)-c_B(r,t)$ to describe the spinodal decomposition of alloys. The evolution of the conservative order parameter $\eta(r,t)$ is given by Equation 1:

$$\frac{\partial\eta(\mathbf{r},t)}{\partial t} = M\nabla^2\left(\frac{\delta F}{\delta\eta}\right) \quad \text{(Eq.1)}$$

where M is the mobility derived from Onsager equation. In the following, we assume that M only depends on the average value of the concentration [21,22]. For an CuAg alloy, this mobility is independent of $\eta(\mathbf{r},t)$ and is equal to 487 s^{-1} out of irradiation at T=769 K. Moreover, only the free energy F depends on $\eta(\mathbf{r},t)$ and can be written as:

$$F(\eta(\mathbf{r},t)) = \int (f(\eta(\mathbf{r},t)) + \frac{\mu}{2}|\nabla\eta(\mathbf{r},t)|^2 dV \quad \text{(Eq.2)}$$

where f is the local coarse grained bulk free energy density. The term $|\nabla\eta(\mathbf{r},t)|^2$ has been added in the free energy to represent the energetic cost associated with interfaces. In order to describe fronts, walls and labyrinthine patterns, the strength of the surface tension is given by the positive constant μ. We assume that f has a double well structure below a critical temperature T_c. All our analysis will be performed for temperatures below T_c. Under this assumption the simplest form of f is given by:

$$f(\eta(\mathbf{r},t)) = \frac{\alpha(T-Tc)}{2}\eta^2 + \frac{\beta}{4}\eta^4 \qquad \text{(Eq.3)}$$

The parameters α and β are positive constants. For a binary alloy, these coefficients can be identified by a comparison with the explicit form of the free energy F. For a CuAg alloy [23], the free energy of this alloy can be written as a regular solid solution out of irradiation. Its ordering energy and its unit cell parameter of the cubic structure are respectively equal to 0.0553 eV and 0.3 nanometer. For a temperature of 769K, coefficients α, β and μ are respectively equal to 1, 0.11 eV/nm^3 and 0.06 eV/nm^5. However, it is more appropriate to think of them as free parameters [2] without any reference to an underlying microscopic model.

The effect of thermal fluctuations can be incorporated in the Cahn Hilliard equation including a noise term $\theta(\mathbf{r},t)$. The resultant model is the well-known Cahn Hilliard Cook model [14]. As the noise at the equilibrium satisfies the fluctuation dissipation relation, we have $<\theta(r,t)> = 0$ and $<\theta(r,t)\ \theta(r',t')> = 2MkT\delta(r-r')\delta(t-t')$, where M is the mobility (1.2 10^{-12} cm^2s^{-1} at T=769 K for CuAg out of irradiation). The bracket <.> denotes the average over the Gaussian noise ensemble. The presence of this noise insures that the system equilibrates to the correct Boltzmann distribution at equilibrium.

Such a model is also referred as the so called B model in the classification of Hohenberg and Halperin in the context of dynamical critical phenomena. The CHC model mimics the time evolution of A-rich and B-rich domains separated by interfaces. Before we proceed, it is relevant to discuss the applicability of the CHC model to real binary alloys at equilibrium. Lattice parameters mismatch in alloys and generate large strain fields. Such strain fields can be easily absorbed modifying the phenomenological coefficients of the bulk free energy density. This is one of success of this equation in material science as first pointed out by Katchaturyan.

Under irradiation, it is possible to add to the CHC equation an athermal diffusion of atoms due to a subcascade formation [11]. In a subcascade, atoms are set in motion during the thermal spike. During the thermal spike, complex defects like voids and dislocations are formed. However these defects do not evolve on the same time scale than the patterning does. These defects assumed to be shrunk are not taken into account to describe the evolution of the microstructure. On the other hand, point defects (vacancies and interstitials of the same species) move rapidly in the solid. It is well-known that these point defects enhance the atomic movement or diffusion at least in alloys and metals. We assume that these point defects only increase the value of the mobility M. For a subcascade of size L (about 10 nanometers), the thermal spike leads to a relocation of A and B atoms according to a simple diffusion equation [7,8]:

$$\frac{\partial \eta}{\partial t} = -W(\eta - p_R * \eta) \qquad \text{(Eq.4)}$$

where W describes the frequency of atom exchange and is a function of the atomic density of the material ρ (100 nm^{-3}), the mixing efficiency of the subcascade ς equal to 1 in the following and the flux of impinging particles ϕ (10^{-4} nm^2s^{-1}) [11]. The number of relocations occurring in a subcascade of size L reduces to ςρ/2 (5 10^4 atoms). The frequency associated with the occurrence of a subcascade at the same position in the solid reduces to ϕL^2 (10^{-2} s^{-1}). The "intensity" of the relocation W in a subcascade is then equal to 500 s^{-1}.

The function $p_R(r)$ is the probability for atoms belonging to the subcascade to be ejected at a distance r from its initial position. Molecular dynamic simulations [15] performed on alloys have shown that this function can be roughly mimicked with an exponential decay $exp(-r/R)$. R defines the spreading of this exchange occurring at the atomic scale (about 0.5 nanometers). The precise form of $p_R(r)$ seems not to play an important role to describe the patterns (see section II). When the parameters R or W tend to zero, the athermal driving term given by Eq.3 plays no role in the evolution of the microstructure. When R tends to infinity, the TDGL equation is similar to the usual equation describing the melting of a copolymer block and characterizes a chemical reversible equation mixing the two compounds [16]. Such an equation (R tends to infinity and W non null) is extensively used to discuss the phase separation in chemically reactive binary mixture. In this case, W is identified with the reaction constant of the reversible equation.

Combining the CHC equation with Eq. 4 leads to the TDLG equation allows studying the microstructure of binary alloys under irradiation. Whereas this equation is not based from first principles, this equation was applied to understand experimental results within a unified framework [2]. Such an equation can be considered as a toy model describing the patterning of materials observed under irradiation [17,18]. The characteristic value of W (500 s^{-1}) is of the same order of magnitude than the mobility M (487 nm^2 s^{-1}) for the CuAg alloy at 769 K. Under irradiation, a balance between the ordering of the alloy driven by the thermodynamic and the disordering induced by the athermal mixing inside the subcascade occurs. This balance may induce a patterning at the nanometric scale, i.e. the characteristic size of a subcascade, in this alloy.

The evolution of this alloy under irradiation is simply given by the TDGL equation. In this study, the initial configuration is given by the random high temperature microstructure quenched below T_c. To visualize this microstructure, The A-rich domains are marked black and B-rich domains are not marked. During the evolution of the alloys under irradiation, the average value of the order parameter $\frac{1}{V}\int \eta(\mathbf{r},t)dV$ is maintained null. Different parameter α, β, μ, M, W and R can be absorbed into the new definition of space and time by introducing rescaled variables for $T<T_c$ following a well-known procedure[19,20].

$$x' = x\sqrt{\frac{\alpha(T_c - T)}{\mu}}$$

$$t' = t\frac{M\alpha^2(T_c - T)^2}{\mu}$$

$$\eta' = \eta\frac{\beta}{\alpha(T_c - T)} \qquad \text{(Eq.5)}$$

$$W' = W\frac{\mu}{M\alpha^2(T_c - T)^2}$$

$$\theta' = \theta\frac{\sqrt{\beta\mu}}{M\alpha^2(T_c - T)^2}$$

θ' is a reduced noise with null mean value and a variance ε equal to $-2kT\Delta\frac{\alpha\sqrt{Tc-T}}{\sqrt{\mu}}$.

The micro structural evolution of the CuAg is then governed by the following equation:

$$\frac{\partial\eta}{\partial t} = \nabla^2(-\nabla^2\eta - \eta + \eta^3) - W(\eta - p_R * \eta) + \theta(r,t) \qquad \text{(Eq.6)}$$

In this equation and in the following, the prime associated with reduced variables was dropped. The values of the order parameters and its second and quadric derivatives are null at the boundaries of the domain described by a subcascade exhibiting a cubic shape.

In previous attempts to catch the main feature of the patterns at low temperature created under irradiation, some authors [17] studied the stability of this equation at high temperature (T above T_c) where the conservative order parameter is null. This implies that the η^3 term is null. It appears then possible to apply the Bloch Floquet theorem to study the stability of different microstructures under irradiation [17]. However, such an analysis is unable to catch the features of the microstructure below T_c since the η^3 term is no more null and becomes the leading term in Eq. 6.

To overcome this difficulty, we use another approach to characterize the microstructure of systems under irradiation. This approach is based on a mean field approximation of the CHC equation [20]. The main interest of this approach is to determine the structure factor S(**k**,t) under irradiation. This function captures all the features of the microstructure. Within the mean field approximation, the effect of the noise on the microstructure has been neglected (ε=0). Following the formalism pointed out by previous authors [20], it is possible to calculate S(**k**,t) under irradiation for long times:

$$S(k,t)k_m(t)^3 = F(k/k_m(t), t\cdot g(WR^4)) \qquad \text{(Eq.7)}$$

Eq. 5 clearly displays that the term $S(k,t)\,k_m(t)^3$ exhibits a Gaussian like shape and is a universal function of k/k_m and WR^4. Depending on the values of R and W, the wave vector $k_m(t)$ defining the patterning exhibits three distinct behaviors. The figure 1 summarizes these conditions:

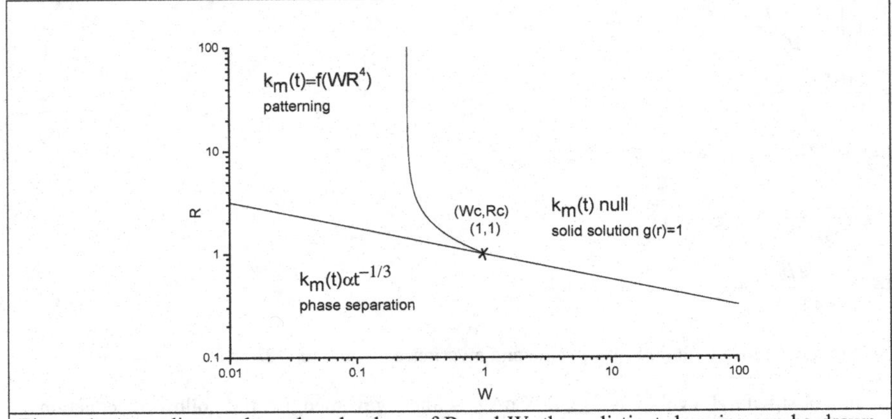

Figure 1: According to the reduced values of R and W, three distinct domains can be drawn. These domains are associated to a phase separation (low W values whatever R values are), a random solid solution (W is important and R above 1), and a steady state. This steady state is defined by a given value of k_m that only depends on W and R. As displayed on the graph, a critical point (Wc,Rc) equal to (1,1) appears.

From Eq. 7, it appears that the structure factor exhibits a Gaussian like shape peaked around a wave vector $k_m(t)$. Depending on different values of W and R displayed on figure 1, three distinct domains can occur under irradiation in a subcascade. For R and W values such that $WR^4<1$, the athermal motion of atoms occurring inside the volume of the subcascade is enable to avoid the demixion of the system. A phase separation occurs and the wave vector of the modulation $k_m(t)$ is a power law of t ($k_m(t) \sim t^{-1/3}$). Increasing the value of W, two distinct domains appear. For large R values ($WR^4>1$), $k_m(t)$ reaches a defined values function on W and R and independent of t. In this domain, a steady state is reached inducing a nano patterning inside the subcascade. For large W and R values, the athermal mixing of atoms resulting from chaotic collisions of atoms inside the subcascade disorder the alloy leading to the creation of a random solid solution. The boundaries layers of three domains intercept at a tri-critical point (Rc,Wc) as displayed on Figure 1. Boundaries of different domains calculated with our mean field approximation have an asymptotic expansion (R tends to infinity) in agreement to expressions obtained by previous authors studying chemical reactions [20]. The previous expressions of boundaries derived from the analysis of the stability [17] of Eq.4 do not satisfy these criteria. Even if the shape of the boundaries associated with three domains and the critical point are different than previous results [17], figure 1 exhibits qualitatively the same features than the one previously calculated [17]. The main interest of our work is to calculate the structure factor.

DIRECT NUMERICAL SIMULATIONS OF THE TDGL EQUATION

In order to assess the validity of the mean field approximation used to calculate S(k,t), we directly compute the structure factor. Whereas the Kinetic Monte Carlo technique is extensively used to study the patterning induced by a subcascade [19], we solved Eq. 4 using a semi implicit finite difference scheme. Since the effect of a subcascade on the atomic concentration is a convolution product (Eq.3), we solve the equation in the Fourier space [25]. Applying an inverse Fourier transform to the Fourier components of the order parameter $\eta(\mathbf{r},t)$, it is possible to determine the patterns induced under irradiation in the real space.

We perform different simulations for different ε values (0.5, 0.01, 0.1, 0) and different R and W values. Results of simulations clearly showed that the thickness of the interfaces depends on ε. Larger ε is, broader the interfaces are. However, these results display that the long time characteristic size of domains does not evolve with ε. This result is in agreement with previous studies [24]. This point assesses the validity of the analysis within the mean field approximation framework. Figure 2 displays the evolution of the microstructure in the real space *for W=0.1* and *R=2.*

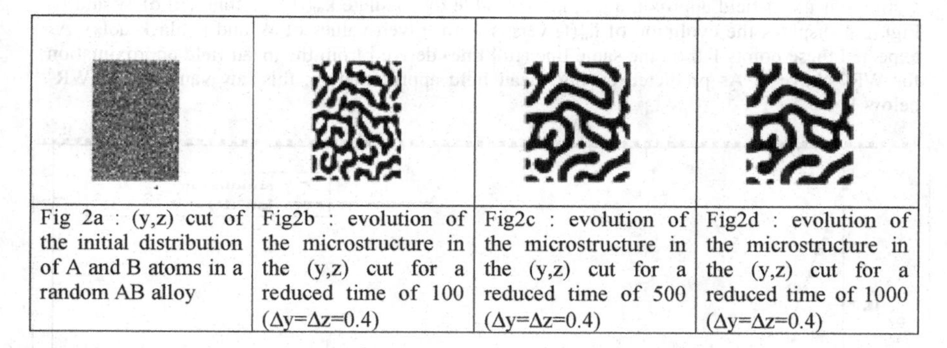

Fig 2a : (y,z) cut of the initial distribution of A and B atoms in a random AB alloy	Fig2b : evolution of the microstructure in the (y,z) cut for a reduced time of 100 ($\Delta y=\Delta z=0.4$)	Fig2c : evolution of the microstructure in the (y,z) cut for a reduced time of 500 ($\Delta y=\Delta z=0.4$)	Fig2d : evolution of the microstructure in the (y,z) cut for a reduced time of 1000 ($\Delta y=\Delta z=0.4$)

From the direct observation of the microstructure driven by irradiation, it appears clearly that a steady state is achieved (see figure 2c and 2d) for these values of W and R. In order to describe the microstructure more quantitatively, Figure 3 displays the evolution of the radial correlation function associated with these four patterns.

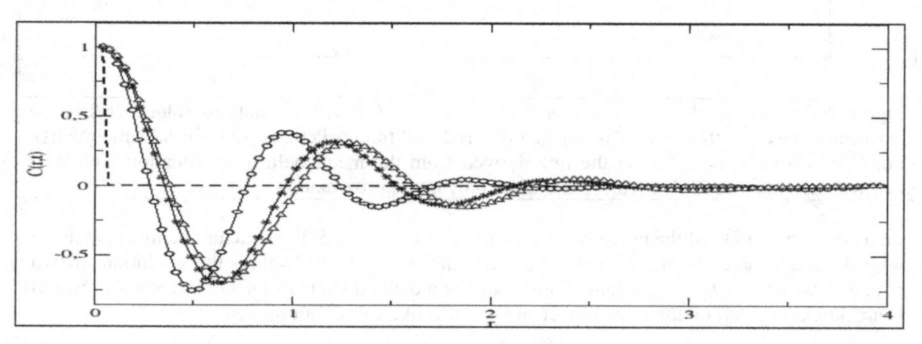

Figure 3. Evolution of the radial correlation function describing the evolution of η(x,y,z,t) in a subcascade under irradiation (W=0.1, R=2). At t=0, this function is a Dirac (dashed line) associated with the quenched random pattern. The radial correlation function exhibits a liquid like behavior (open dots). Increasing the time, oscillations associated with characteristic lengths appear around the reduced distance r=1and r=2 (open triangles and stars). For long times, this function no more evolves insuring the existence of a steady state for these values of R and W.

As the asymptotic expansion of boundaries displayed on Figure 1 are similar to those derived previously [20], it seems that the mean field approximation can be safely applied to capture the features of Eq. 4. However, to assess this point directly, the variation of k_m as a function of R is plotted on figure 4 for a given value of W for a long simulation time, when the steady state is achieved. To be sure the steady state is achieved; two distinct simulations using as initial configuration a random and a perfectly ordered state were performed. We check that the structure factors associated with two final states are similar; assessing the steady state is reached. Within our mean field approximation, it is possible to calculate $k_m(t)$ as a function of W and R. Figure 4 displays the evolution of $k_m(t)$ versus R for given values of W and t (black dots). As expected these points fall on the same line (full line) derived from the mean field approximation for WR^4 above 1. As predicted by our mean field approximation, this law vanishes for WR^4 below 1.

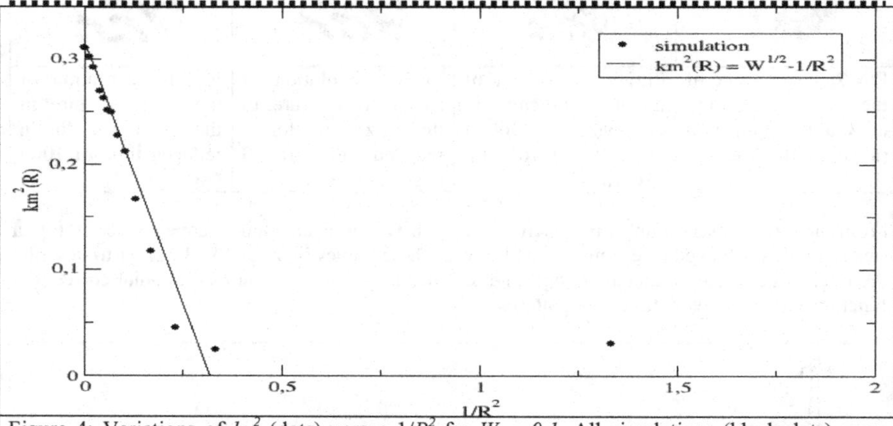

Figure 4: Variations of k_m^2 (dots) versus $1/R^2$ for $W = 0.1$. All simulations (black dots) were performed once the steady state is reached (the reduced time is kept to 2000 for all simulations). For WR^4 above 1, dots fall on the line derived from the mean field approximation. For WR^4 below 1, $k_m(t)$ do not evolve with R as expected by our approximation.

To assess the validity of the universal feature of $S(k,t)k_m^3$ in Eq.5, the structure factors associated with different values of W, R and t were calculated. Figure 5 displays the evolution of two structure factors $S(k,t)k_m^3$ as a function of k/km for a defined value of $tg(WR^4)$. Results extracted from simulations fall on the same universal Gaussian like curve (not plotted).

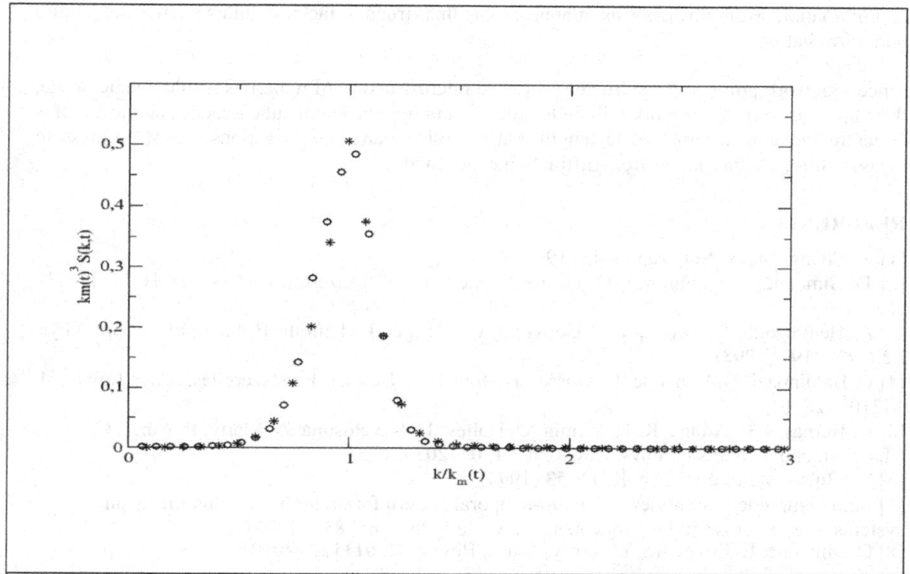

Figure 5: Plot of the normalized structure function as a function of k/k_m for different values of W, R and time extracted from numerical simulations. For a given value of $tg(WR^4)$, these two curves exhibit the same behavior as a function of k/k_m assessing the universal feature of $S(k,t)$.

The analysis of direct simulations of Eq.4 seems then to be accurately described within a mean field approximation framework as firstly pointed out by previous authors who have studied the structural evolution of diblocks copolymers out of irradiation [20].

DISCUSSION AND CONCLUSION

Since about fifty years, experimental works based on TEM or grazing incidence X ray diffraction have clearly shown that alloys and ceramics exhibit a nano pattering under irradiation [1,2,3]. Many works were devoted to study the nano patterning induced by ion beam mixing in solids [17,18,19]. Understanding the nano patterning will help to tailor materials with well-defined microstructures at the nanometric scale. The slowing down of impinging particles in solids leads to a complex distribution of subcascades. Moreover, each subcascade will give rise to an athermal diffusion of atoms inside the subcascade. In this work, we focused our attention on this last point. Based on the well-known Cahn Hilliard Cook equation, we analytically calculate the structure factor describing the nano patterning induced by irradiation inside a subcascade within the mean field approximation framework. We show that this analytical structure factor mimics the structure factor extracted from direct numerical simulations assessing the accuracy of our approximation. From this analysis, it appears that this structure factor exhibits a universal feature under irradiation.

Since this work points out modifications of the microstructure of materials inside a subcascade, the following part of our work will be to study if and how different subcascades can modify this structure factor at a larger scale length. On the other hand, investigations are in progress to measure this structure factor from diffraction experiments.

REFERENCES

[1] Y. Cheng, Mater. Sci. Rep. **5**, 45 (1990).
[2] D. Simeone, G. Baldinozzi, D. Gosset, S. Le Caer, L. Mazerolles, Phys Rev B 70, 134116 (2004).
[3] A. Benyagoub, L. Levesque, F. Couvreur, C. Mougel, C. Dufour, E. Paummier, Appl. Phys. Lett. 77, 3197 (1998).
[4] G. Baldinozzi, D. Simeone, D. Gosset, I. Monnet, S. Le caer, L. Mazerolles, Phys Rev B 74, 132107 (2006).
[5] H. Bernas, J. P. Attane, K. H. Heinig, D. Halley, D. Ravelosona, A. Marty, P. Auric, C. Chappert, and Y. Samson, Phys. Rev. Lett. **91**, 077203 (2003).
[6] W. Bolse, Mater. Sci. Eng. R. **12**, 53 (1994).
[7] for a comprehensive review of spatio-temporal pattern formation in reaction diffusion systems, see M. Cross and P. Hohenberg, Rev Mod. Phys. 65, 851 (1993).
[8] D. Simeone, L. Luneville, Y. Serruys, Phys. Rev. E **82**, 011122 (2010).
[9] G. Martin, Phys. Rev. B 30, 1424 (1984).
[10] G. Martin, P. Bellon, Solid State Phys. **53-54**, 1 (1997).
[11] D. Simeone, L. Luneville, Phys. Rev. E 81, 21115, (2010).
[12] D. Simeone, C. Dodane, D. Gosset, P. Daniel, M. Beauvy, Journal of Nucl. Mat. 300, 151, (2002).
[13] H. Cook, Brownian motion in spinodal decomposition, Acta Metal. 18, 297, (1970).
[14] D Fisher, D. Huse, Phys. Rev. B 38, 373 , (1988).
[15] R Enrique, K. Nordlung, R . Averbach, P. Bellon, Journal of Applied physics 93(5), 2917, (2003).
[16]S. Glotzer, D. Stauffer, N Jan, Phys. Rev. Lett. 72,4109 (1994) .
[17] R. Enrique, P. Bellon, Phys. Rev. Lett. 84(13), 2885, (2000).
[18] R. Enrique, P. Bellon, Phys. Rev. B 63, 134111, (2001).
[19] J. Ye, P. Bellon, Phys. Rev. B 73, 224121 (2006).
[20] S. Glotzer, A. Coniglio, Phys. Rev. E 50(5), 4241 (1994).
[21] G. Martin , Phys. Rev. B 41, 2279, (1990).
[22] J.F. Gouyet, Phys. Rev. E, 51(3), 1695, (1995).
[23] L. Wei, R. Averback, J of Appl. Phys. 81, 613 (1997).
[24] T. Rogers, K. Eldere, R. Desai, Phys. Rev. B 37(16), 9638 (1988).
[25] J. Zhu, L. Cheng, J. Shen, V. Tikare, Phys. Rev. E 60(4), 3564, (1999).

Radiation Effects

Mater. Res. Soc. Symp. Proc. Vol. 1514 © 2013 Materials Research Society
DOI: 10.1557/opl.2013.196

Nanostructuration of Cr/Si layers induced by ion beam mixing

L. Luneville[1], L. Largeau[2], C. Deranlot[3], N. Moncoffre[4], Y. Serruys[5], F. Ott[6], G. Baldinozzi[7], D. Simeone[8]

[1]DEN/DANS/DM2S/SERMA/LLPR/LRC-CARMEN, CEA Saclay, 91191 Gif-sur-Yvette, France
[2]LPN-UPR20/CNRS, Route de Nozay, 91460 Marcoussis, France
[3]Unité Mixte de Physique CNRS/Thales, 1 Avenue Augustin Fresnel, 91767 Palaiseau, France
[4]IPNL-IN2P3, 69622 Villeurbanne, France
[5]DEN/DANS/DMN/SRMP, CEA Saclay, 91191 Gif-sur-Yvette, France
[6]DSM/IRAMIS/LLB, CEA Saclay, 91191 Gif-sur-Yvette, France
[7]CNRS-SPMS/UMR 8580/ LRC CARMEN Ecole Centrale Paris, 92295 Châtenay-Malabry, France
[8]DEN/DANS/DMN/SRMA/LA2M/LRC-CARMEN, CEA Saclay, 91191 Gif-sur-Yvette, France

ABSTRACT

This work clearly demonstrates that the X Ray Reflectometry technique (XRR), extensively used to assess the quality of microelectronic devices can be a useful tool to study the first stages of ion beam mixing. This technique allows measuring the evolution of the Si concentration profile in irradiated Cr/Si layers. From the analysis of the XRR profiles, it clearly appears that the Si profile cannot be described by a simple error function.

INTRODUCTION

Thin films play a dominant role in modern technology instigating a great deal of research interest from the perspective of basic science. Extensive work has been carried out on thin transition metal films (Fe, Co, Cr, ...) on Si wafers to understand the formation mechanism of various silicide compounds by thermally induced reaction at the Si/Metal interface[1-3]. In this context, ion beam mixing has an advantage; the ballistic nature of ion beam mixing at low temperature makes possible to mix two distinct elements even in an immiscible system [4]. Although ion beam mixing allows overcoming the equilibrium phase diagram, physical mechanisms responsible for the mixing are far from being clearly understood. Despite conventional methods to obtain information on ion beam mixing include Rutherford Backscattering Spectrometry (RBS), secondary ion mass spectrometry and Auger electron spectrometry with ion etching, these techniques are not able to probe nanometric layers.

By contrast, the X-Ray Reflectometry (XRR) technique is well known as a characterization tool for studying stratified media like semi conductor super lattices [5] and Langmuir Blodgett films. In XRR, the intensity of an X-ray beam is specularly reflected on the sample. Due to high precision diffractometers, this technique offers advantages including high spatial sensitivity at the nanometric scale, high penetration and non destructive capability. At low incidence angles, XRR measurements are sensitive to modulation profiles of electronic density and absorption coefficient in a layered structure. The main feature of this technique remains the fact that the resolution in depth does not depend on the material but only on the detector resolution. On the other hand, many models, based on the Parrat [6] formalism, are now available to simulate XRR diagrams over a large angular range. Comparison of simulated and experimental diagrams can then give information on the atomic composition of layers at the nanometric scale [see 7 for a comprehensive approach of XRR]. By comparing XRR diagrams calculated with experimental data, quantitative information about interface modification upon ion irradiation can be achieved.

The goal of this work is to have a quantitative characterization of the structural evolution of the Cr/Si layer under a wide range of ion irradiation fluences. In this paper, XRR measurements as well as Scanning Transmission Electronic Microscopy and Energy Dispersive X-ray analysis (STEM-EDX) were performed to estimate the evolution of the chromium and silicon concentration profiles as well as the microstructure of the interface versus fluences, at the nanometric scale. The ultimate goal of this work is to provide experimental data to understand the underlying mechanisms responsible for the ballistic mixing in materials.

EXPERIMENT

Choice of the material and the temperature

Several models were developed to describe ion beam mixing at the interface of bilayers. Some authors claim that ion beam mixing results from multiple collisions between recoils atoms in materials with an average atomic number below 20 [8]. Above this critical value, the mixing results from a Brownian motion of atoms in sub cascades. This last mixing mechanism remains difficult to understand. For the Cr/Si system studied in this work, the average atomic number is equal to 19 and the ballistic mixing mainly results from separated collisions [9].

A large number of studies on ion beam induced mixing have been carried out to understand transport mechanisms and phase formation processes in metal/semi conductors [10,11,12]. Depending on the temperature, two distinct regimes occur. At low temperature, the mixing seems to be trigged by a temperature independent mixing regime. Above a critical temperature, it appears that the amount of mixing depends on the temperature. To avoid such a thermally assisted ion mixing, we have then irradiated Cr/Si layers at room temperature (300K) well below the expected critical temperature for this system (estimated lying between 415 and 570 K [11]).

The Pulsed Vapor Deposition (PVD) technique was used to elaborate 20 nanometers Cr layers on (100)Si wafers.

Choice of the irradiation

The profiles of different ions set in motion during the ballistic collision phase induced by 80 keV Kr ions and the energy deposited by these atoms at the interface were calculated within the Binary Collision Approximation (BCA) using the SRIM-2006 code (monolayer collision option). Figure 1a displays the penetration depth of Kr incident ions (full line) and the distribution of recoils atoms set in motion by Kr incident ions (dotted line). As the number of recoils is three orders of magnitude more important than the number of Kr incident ions, the ballistic mixing is mainly induced by recoil atoms. Since the incident energy of Kr ion is equal to 80 keV, the energy transferred to atoms in the layer is mainly due to atomic collisions. The effect of the electronic stopping power can then be neglected. The nature and the energy of the incident ion were chosen to maximize the ballistic mixing at the Cr/Si interface. To assess this point, the first derivative of the deposited energy as a function of the depth was calculated by SRIM-2006 and is plotted figure 1b. The maximum energy loss due to nuclear collision occurs at the Cr/Si interface as expected. From Figure 1, the density of deposited energy in nuclear collision F_d at the interface (at a depth of 20 nanometers as shown on figure 1) is kept constant and equal to 0.8 KeV/nm for all the irradiations.

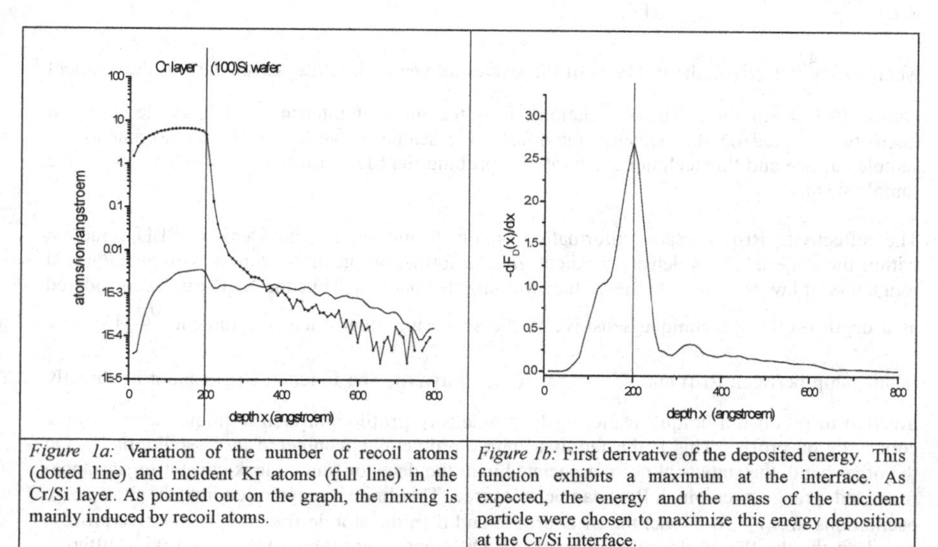

Figure 1a: Variation of the number of recoil atoms (dotted line) and incident Kr atoms (full line) in the Cr/Si layer. As pointed out on the graph, the mixing is mainly induced by recoil atoms.

Figure 1b: First derivative of the deposited energy. This function exhibits a maximum at the interface. As expected, the energy and the mass of the incident particle were chosen to maximize this energy deposition at the Cr/Si interface.

In order to compare experimental results with theoretical assumptions and previous results [11,12,13,14], six different Cr/Si layers were irradiated with 80 keV Kr ions at room temperature, using a defined flux of 7.7 10^{12} $cm^{-2}s^{-1}$. The evolution of the concentration profile and the microstructure were studied over a large fluences range (from 5 10^{14} to 2 10^{16} cm^{-2}).

Choice of the experimental technique

The XRR is nowadays a well established technique for investigating surface and buried interfaces. The main part of its interest comes from the opportunity to study these buried interfaces with high spatial resolution (nanometer length scale). The specular reflectivity of an interface is determined by changes in the scattering length density (number of electrons per nanometer)

$$\rho = dr_e \sum_i c_i Z_i \qquad \text{(Eq.1)}$$

(where r_e is the Lorentz classical radius of the electron (2.818 10^{-15}m), d is the atomic density, Z_i and c_i are respectively the atomic number and the atomic fraction of the element i) in the direction normal to the surface. In XRR, the normalized intensity of the reflected X-ray beam, R(q), is measured as a function of the detector angle 2θ. This angle defines the angle between the incident and the scattered wave vectors associated with the incident and scattering X-ray beam. Introducing the scattering vector **q** lying in the plane defined by the normal **Z** to the surface and the wave vector associated with the incident X ray beam $\mathbf{k}_{inc}$, it is possible to plot the evolution of R(q):

$$R(q)=\frac{I_R(q)}{I_0} \qquad \text{(Eq.2)}$$

where $q=\frac{4\pi}{\lambda}\sin(\theta)$ is the modulus of the scattering vector, λ is the wavelength of the incident photon (0.154 nm for a CuKα radiation), I_0 is the incident intensity and I_R is the reflected intensity collected on the detector. Obviously, the scattering vector is always normal to the sample surface and this technique then allows probing the SDL of different layers normal to the sample surface.

The reflectivity R(q) contains information on the Scattering Length Density (SLD) gradient within the material. This density gradient gives information on the thickness, composition and roughness of layers parallel to the surface forming the material. Thus the XRR can be considered as a depth profiling technique sensitive to the scattering length density gradient $\frac{d\rho}{dz}$ [13]. The relationship between R(q) and $\frac{d\rho}{dz}$ is not straightforward. The function R(q) cannot be directly inverted to produce a unique scattering length density profile and then a unique set of atomic concentration profile. This is the familiar "phase problem" encountered in crystallography. On the other hand, the statistical noise associated with the detector (due to leak current for instance) restricted q range over which R(q) can be measured. This finite q range often induces short range oscillations in $\rho(z)$. To extract the SLD profile and then the atomic fraction, an iterative process based on the the Parrat algorithm is used [6]. Different layers (about ten layers) with different thicknesses and SLD were used to simulate the evolution of the experimental reflectivity curves at different fluences. The comparison between the model and the measured reflectivity is achieved by calculating a χ^2 reliability factor (the sum of squares of the weighted deviations). The thickness of each layer is derived from the distances between maxima of the Keissig fringes of the reflectivity curve, the amplitudes of the different oscillations allow fitting values of the SLD, whereas the roughness of each layer can be extracted from the slope of the reflectivity curve. Only six free parameters are needed to fit the reflectivity curve for the non irradiated sample as displayed on figure 2a. This number of free parameters reaches 15 for the most irradiated sample. For all refinements, the reliability factor χ^2 remains below 2. The comparison between fitted and experimental reflectivity curves are in very good agreement. From the knowledge of the atomic density of different phases associated to the Cr/Si system out of irradiation, it was possible to extract the Si and Cr atomic fractions from the fitted SLD.

To assess the validity of the Si and Cr profiles extracted from different fits, different transverse cross sections on irradiated samples have been done by focused ion beam (FIB). Different high angle annular dark field pictures (STEM-HAADF) were then collected as a function of the fluence. This technique is sensitive to the variation of the atomic number Z. The high Z atoms appear brighter than the light atoms (low Z values). Such dark field pictures allow a direct evolution of the interface versus the ion fluences. From this picture, it becomes possible to study the broadening of the interfaces induced by irradiation. On the other hand, the Cr/Si ratio has also been measured along the ion beam incident direction (the **Z** direction) by Energy Dispersive X-Ray analysis (EDX). This technique insures direct measurements of the atomic fraction of these two elements. This direct measure of the Si and Cr profiles, independent of the XRR technique, then assesses the model used to fit XRR data.

RESULTS AND DISCUSSION

Quality of XRR data

To check the quality of XRR data, we firstly measure the XRR diagrams on pristine Cr/Si layers. A 0.1 mm Ni filter was used to select only the CuKα_1 radiation. A Goebbel mirror insures the incident X ray beam to be parallel (its divergence is inferior to 0.01°) [10]. The footprint of the X-ray beam at low 2θ values (below 0.2°) was used to calibrate the 2θ scale. The accuracy of the angles was then inferior to 0.01°. Off specular measurements with a 2θ offset of 0.3° was used to correct the reflectivity from background. Thanks to this procedure, it was possible to measure a reflectivity curve over 8 orders of magnitude.

Figure 2a displays the XRR diagram of a pristine Cr/Si sample. From this diagram, it is possible to extract the concentration of Si. Such a concentration is plotted as a function of the depth in figure 2b. The thickness of the Cr layer is equal to 20 nanometers as expected and the roughness of the Cr/Si layer is equal to 2 nanometers in fair agreement with AFM measurements.

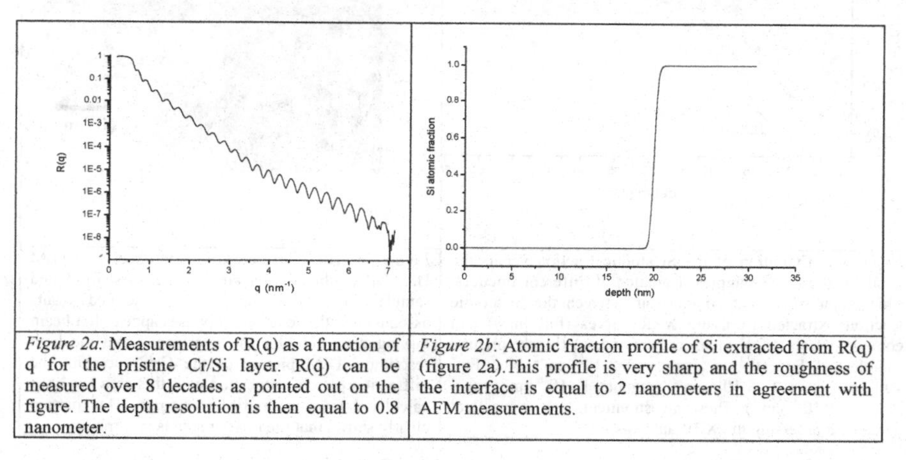

Figure 2a: Measurements of R(q) as a function of q for the pristine Cr/Si layer. R(q) can be measured over 8 decades as pointed out on the figure. The depth resolution is then equal to 0.8 nanometer.

Figure 2b: Atomic fraction profile of Si extracted from R(q) (figure 2a).This profile is very sharp and the roughness of the interface is equal to 2 nanometers in agreement with AFM measurements.

Analysis of concentration profiles of irradiated samples

Figure 3a displays the evolution of the silicon atomic fraction as a function of the depth in the material for different fluences. The effect of the ballistic mixing appears clearly on these graphs. The Si profiles extracted from the analysis of the XRR curves agree quite well with the measured profiles derived from the EDX analysis.

The shape of the Si profiles between the pristine and the irradiated profiles are quite different. The ion beam mixing leads to a shift of the Si profile inside the material. This point is the signature that the diffusion of Cr and Si induced by the ballistic collision are quite different at least for the step of the ion beam mixing.
Analysis of the Si profiles extracted from XRR curves shows that these profiles cannot be described by a single error function, as clearly displayed on figure 3a. We observe two distinct plateaux for Si concentration equal to 0.33 and 0.67 respectively. So, three distinct error functions are needed to fit experimental data. The thicknesses of the three distinct domains, a few nanometers, are extracted from the error functions and increase with the fluence. This result points out the main interest of the XXR technique to describe ion beam mixing. From the high accuracy of the XRR technique, it is possible to show that ion beam mixing in low Z materials is a complex mechanism which cannot be resumed as a single diffusion process. This result disagrees with previous results based mainly on RBS measurements [11,12,15].

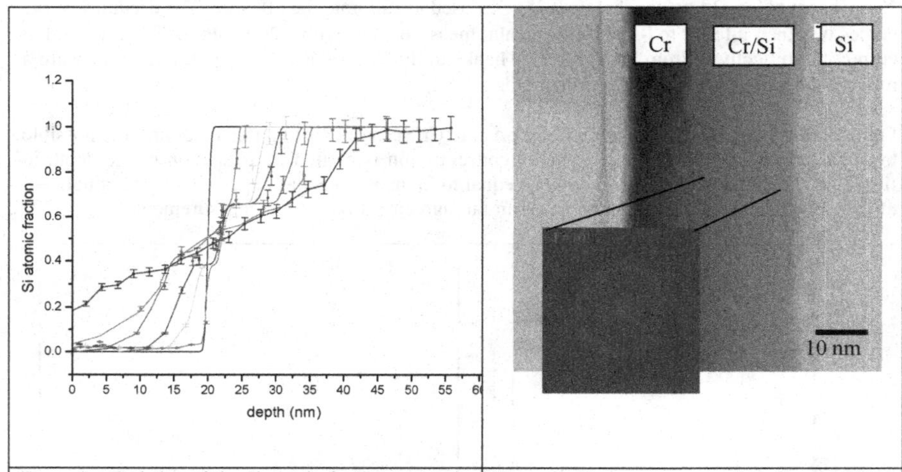

Figure 3a: Evolution of the Si atomic fraction versus the depth of the Cr/Si samples irradiated at different fluences. This graph shows the comparison between the Si atomic fraction extracted from the XRR curves (full lines) and computed from EDX measurements (dots) (black: pristine sample, red: F=5 10^{14} cm^{-2}; green: F=2 10^{15} cm^{-2}, blue: F=5 10^{15} cm^{-2}, navy: F=8 10^{15} cm^{-2}, magenta: F=10^{16} cm^{-2}; royal blue: F=2 10^{16} cm^{-2}). These measurements are in excellent agreement assessing the XRR analysis.

Figure 3b: Transverse cross section observed STEM HAADF of the mixing area of the most irradiated sample. From this picture, well defined fronts associated with the different phases appear. Ion beam mixing does not blur the interface. From the high resolution TEM picture of the Cr/Si domain, the diffraction pattern associated with this area has been calculated from a Fourier transform. This pattern clearly shows that the mixing area is amorphous.

Figure 3b displays the high angle annular dark field pictures (STEM-HAADF) associated with the transverse cross section of the most irradiated sample. The analysis of the HAADF-STEM picture clearly shows that the interfaces between the Cr rich, Cr/Si and Si rich layers remain sharp even at high fluence. On the other hand, the Fourier transform of the high resolution TEM picture of the Cr/Si area clearly points out that this area is amorphous. This result assesses previous investigations on silicides produced by ion beam mixing [14].

From the knowledge of the Si profiles, it is possible to accurately compute the evolution of the mixing size, calculated as the distance between the two inflexion points of the two extreme error functions, versus the fluence. Figure 4 displays the evolution of this mixing length L for different fluences. It appears clearly that this length is a square root function of the fluence. This result disagrees with previous investigations [11,12,13] that claim a linear relation between the mixing length and the fluence. Based on the seminal work of Sigmund [9], the expected value of the mixing length versus the fluence has been plotted (full line) for the irradiations. As pointed out by many authors [15], it appears that the mixing length is one order of magnitude higher than the expected value (full line) derived from the Sigmund formalism [9].

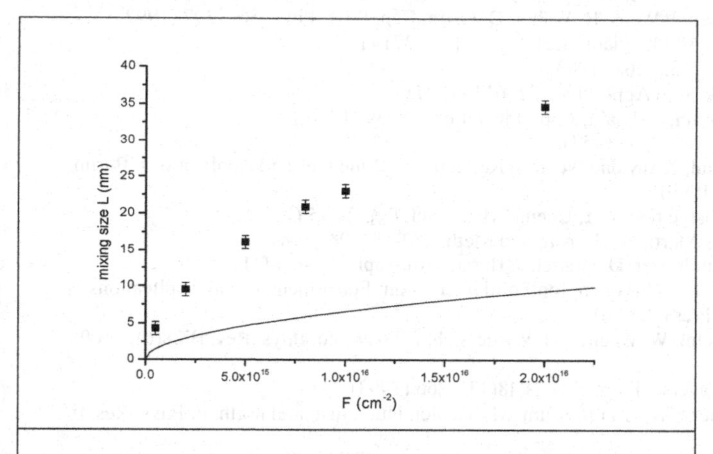

Figure 4: Variation of the mixing length as a function of the fluence derived from the Si profile. The comparison between the mixing length derived from XRR curves (dots) and calculated from the Sigmund formalism [9] (line) clearly shows a large discrepancy (an order of magnitude).

Such a result is not surprising as we have demonstrated that the formation of the Cr/Si compounds under ion beam mixing seems to be a complex mechanism. A simple approach based either on a relocation of displaced atoms like in the Sigmund formalism or the mixing of atoms in a displacement cascade cannot handle the appearance of an amorphous domain nor the appearance of defined compounds.

CONCLUSION

This work demonstrates that the XRR technique can be a useful tool to follow the first stages of mixing induced by irradiation. Whereas many parameters are needed to extract a concentration profile from XRR curves, comparison of profiles extracted from XRR curves and direct EDX measurements clearly points out that the Si profiles derived from XRR curves are accurately measured. From this analysis, it appears that the Si profile cannot be fitted by a single error function at least for the first stage of the mixing. From the TEM analysis of the transverse cross section, we demonstrate that the mixing area is amorphous for each fluence.

It appears from this work that mechanisms associated with ion beam mixing are far from being clearly understood even for a simple and well studied Cr/Si system. The underlying thermodynamics or kinetics responsible for the formation of the amorphous Cr/Si mixing layer remains unclear.

REFERENCES

[1] A. Reader, A. Ommen, P Weijs, R. Walter, D. Ostra, Rep. Prog. Phys. 56, 1397 (1992).
[2] C. Calandra, O. Bisi, G. Ottaviani, Surf. Sci. Rep. 4, 271 (1985).
[3] G. Rubloff, Surf. Sci. 132, 268 (1983).
[4] L. Wei, R. Averback, J. of Appl. Phys. 81, 613 (1997).
[5] L. Chang, A. Segmuller, L. Esaki, Appl. Phys. Lett. 28, 39 (1976).
[6] L. Parrat, Phys. Rev. 95, 359 (1954).
[7] J. Daillant, A. Gibaud, X-ray and Neutron Reflectivity: Principles and Applications. Berlin, Heidelberg: Springer. (1999).
[8] see for a comprehensive review Y. Cheng, Mater. Sci. Rep. **5**, 45 (1990).
[9] P. Sigmund, A. Gras Marti, Nucl. Instr. and Meth. 182-183, 25 (1981).
[10] D. Simeone, G. Baldinozzi, D. Gosset, J. Berar, J. of Appl. Cryst. 44, 1205 (2011).
[11] A. Nastasi, J. Mayer, J. Hirvonen, Ion Solid interaction: Fundamentals and Applications, Cambridge University Press, (1996).
[12] R. Averback, L. Rehn, W. Wagner, H. Wiedersich, P. Okamoto, Phys. Rev. B 28(6), 3100 (1983).
[13] J. Desimoni, A. Traverse, Phys. Rev. B 48(13), 266 (1993).
[14] W. Johnson, Y. Cheng, M. Van Rossum, M. Nicolet, Nucl. Inst. and Meth. in Phys. Res. B 7/8, 657 (1985).
[15] S. Tobbeche, A Boukhari, R. Khalfaoui, A. Amokrane, C. Benazzouz, A. Guittoum, Nucl. Instr. and Method in Phys. Res. B 269(24), 3242 (2011).

Mater. Res. Soc. Symp. Proc. Vol. 1514 © 2013 Materials Research Society
DOI: 10.1557/opl.2013.386

Radiation damages on mesoporous silica thin films and bulk materials

X. Deschanels[1], S. Dourdain[1], C. Rey[1], G. Toquer[1], A. Grandjean[1], S. Pellet-Rostaing[1] and O. Dugne[2] and C. Grygiel[3] and F. Duval[4] and Y. Serruys[5]

[1]Institut de Chimie Séparative de Marcoule, UMR5257 CEA-CNRS-UM2-ENSCM, F-30207 Bagnols sur Cèze, France.

[2]CEA Marcoule-DTEC, F-30207 Bagnols-sur-Cèze, France.

[3]CIMAP, CEA-CNRS-ENSICAEN-UCBN, F-14070 Caen Cedex 5, France.

[4]CNRS, CEMHTI, UPR3079, F-45071 Orleans 2.

[5]CEA, DEN, Serv Rech Met Phys, Lab JANNUS, F-91191 Gif Sur Yvette, France.

ABSTRACT

Mesoporous silicas are highly potential materials for applications in the nuclear field like separation, recycling or nuclear wastes confinement. In this work, the effects of the radiation damage on the mesoporous network were investigated by XRR (X-Rays Reflectivity) and nitrogen adsorption isotherm on respectively mesoporous organized thin films (SBA) and disordered bulk mesoporous materials (Vycor glass). The article attempts to answer the question of the existence of a relationship between the rigidity of the mesoporous silica network, and the behavior of silica materials under irradiation.

INTRODUCTION

Two main solutions are usually proposed for conditioning nuclear wastes. The first one, used for low and intermediate level activity wastes, consists in embedding the contaminated particles or species in a matrix (hull compaction process, cementation process, bitumen process…). In the second concept used for the immobilization of high level waste (HLW) arising from the reprocessing of the spent fuel, radionuclides are incorporated in the network of mineral matrices such as ceramics or glasses. Thanks to their random network, borosilicate glasses are able to accommodate a wide range of extremely complex waste stream compositions (containing up to 30 different elements). This matrix is obtained by a melting process at a relatively high temperature (close to 1200°C). This high temperature process can be problematic in case of volatile species such as iodine or caesium. It is thus of interest to develop a new, low temperature process, to incorporate mobile species into inorganic matrices. Over the past two decades, the development of mesoporous materials provided a number of new possibilities for encapsulating nuclear wastes [1-2]. Mesoporous silicates are typically synthesized under mild or hydrothermal conditions using a structure-directing agent or template that can be removed subsequently, leaving a material with void spaces. These soft synthetic routes are a first determining advantage of mesoporous solids as conditioning matrices for volatile species (ie, I and Cs). A second advantage is that their porosity can be closed under relatively soft conditions like mild thermal treatment, or suited chemical stresses, in order to ensure a durable confinement. This ability of mesoporous structure to collapse under given stresses is thus of great advantage regarding the foreseen applications, but it also raises the question of the behaviour of such materials under self-irradiation. Klaumünzer [3] observed the shrinkage of mesoporous Vycor glasses irradiated by various ions having high electronic stopping power (S_e) (^{40}Ar-70MeV, ^{86}Kr-260MeV, ^{129}Xe-340MeV), under fluences lower than 10^{14} cm^{-2}. This behaviour was attributed to the formation of amorphous tracks. In vitreous silica, at room temperature, the threshold for track formation is observed in the range of values 2 to 4 keV/nm [4]. Silica gels with non-organised nanopores were also irradiated with He ions (E=3.6 MeV), for S_e lower than threshold of track formation, at fluences ranging from 10^{15} to $5x10^{15}$ He.cm^{-2} [5]. Even in these soft irradiation conditions, the authors observed the densification and consolidation of the material after irradiation.

The objective of the study is to investigate the effects of the radiation damages onto mesoporous silica structure and a potential radiation induced collapse. In order to investigate the effect of the rigidity of the mesoporous silica structures, two different materials: Vycor glass samples and mesoporous silica gel thin films were irradiated by ions (He, Ar). Such particles have different electronic stopping power, taken lower than the threshold of track formation, in these experiments.

EXPERIMENT

Samples preparation

Mesoporous thin films were obtained by classical sol-gel process [6]. The sol was prepared in two steps. In the first step a stock solution of pre-hydrolysed silica precursor was obtained by refluxing TEOS (TertaEthylOrthoSilicate), millipore water and hydrochloric acid. In the second step the surfactant dissolved in ethanol in acidic condition was added to the first solution. The mesoporous films were templated by the pluronic surfactant P123 to produce the SBA15 structure. The second solution was aged for 2 hours, before adding, 24g of H_2O/HCl solution (pH = 1.25). The thin film was then made by dip-coating at a constant withdrawal velocity of 14 cm/min on a silicon substrate in the final sol of molar composition $1TEOS:72C_2H_5OH:8H_20:3HCl:0.012P123$. Such a composition was known to yield films having the 2D hexagonal p6m symmetry; with typically 100 nm average thickness and cylindrical pores having a diameter of 5.3 nm [7]. The film was thoroughly rinsed with ethanol at 80°C for 2h to remove the surfactant so as to produce a mesoporous film.

Vycor glasses were supplied by "Vitrabio" (Germany). This glass is a silicate glass (98% SiO_2) with numerous tiny pores having typical radii of about 3 nm. The average mass density of bulk Vycor is 1.7 g.cm^{-3}. The samples had typical size of 20x8x0.2 mm^3. The sample sizes were chosen to be compatible with the sensitivity of the device used for the measurement of the N_2 adsorption-desorption isotherm.

The samples were irradiated at room temperature at JANNUS (CEA Saclay, France) and CEMHTI (Orléans, France) facilities for thin film, at GANIL (Caen, France) facility for bulk Vycor glasses. Some irradiations parameters are reported in the table I. In order to investigate the effects of the electronic stopping power S_e, He and Ar particles were bombarded onto the samples. The energies of the ions were chosen to obtain a homogeneous value of S_e into the volume of the sample.

The mesoporous thin films were characterized before and after irradiation by X-Rays Reflectometry (XRR) performed on a Brucker D8 Advance diffractometer. The surface areas and the pore size distribution of the Vycor glasses were obtained by ASAP 2020, using nitrogen gas as adsorptive at 77 K. Before these measurements, the samples were degassed at 350 °C for 4 h under a vacuum of 10^{-2} mbar. The morphology of all the samples was also investigated by Scanning Electron Microscopy (SEM) using a ZEISS Supra 55 microscope.

Table I. Irradiation conditions (SRIM 2012 calculations).

Thin mesoporous SiO_2 films on Si substrate e_{SiO2}~100nm					
Ions	**Φ (cm^{-2})**	**E (MeV)**	**S_e (keV/nm)**	**S_n (keV/nm)**	**Ions range (µm)**
Ar	0.05-1 x 10^{15}	20	3.4	0.012	~7
He	0.05-1 x 10^{16}	0.5	0.26	5.9 x 10^{-4}	~2
Vycor glasses e~200µm					
Ions	**Φ (cm^{-2})**	**E (MeV)**	**Se (keV/nm)**	**Sn (keV/nm)**	**Ions range (µm)**
Ar	0.03-2 x 10^{14}	548	1.5	1 x10^{-3}	~250
He	0.005-1 x 10^{16}	32	0.03	1.6 x 10^{-5}	~615

Characterizations of the Vycor glasses

The BJH [8] method was used to calculate the pore diameter distribution of Vycor glasses, from the adsorption branch of the isotherm. This distribution, for the sample irradiated by helium at the highest fluence, is shown in figure 1a before and after irradiation. There is a slight shift towards the larger diameters after irradiation. This phenomenon was observed for all samples, which results in an increase in the average pore diameter as a function of fluence, as shown in figure 1b. The increase in the average diameter of the distribution is at most 20%. Together with the increase of the mean diameter, there was a slight decrease in the specific surface area of the glasses up to 25% (figure 1c). In agreement with these results, SEM observations of these samples do not demonstrate significant changes of the meso-structure of the samples.

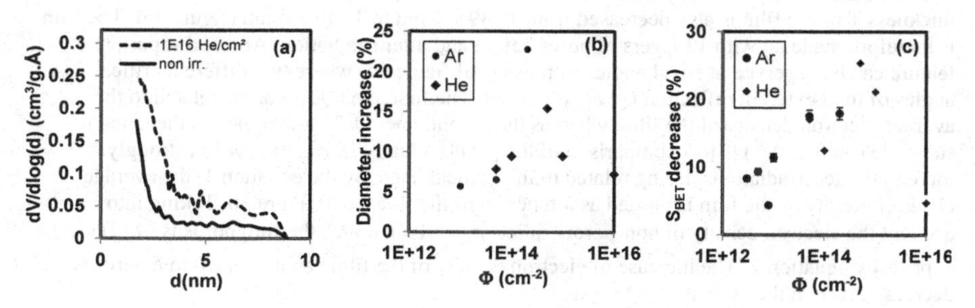

Figure 1. Evolution of the characteristics of Vycor glasses under irradiation, a) Pore diameter distribution for 32 MeV-He, b) and c) Pore diameter and specific surface area versus fluence for 32MeV-He and 548MeV-Ar.

Characterizations of the mesoporous thin films

The XRR curves exhibit Kiessig oscillations [9] (better visible between 0.04 and 0.075 Å^{-1}) that are characteristic of a film having a homogeneous total thickness T (figure 2). The periodicity d of the stacking of pores produces Bragg reflections, located for example at Q=0.085, 0.160 and 0.244 $Å^{-1}$ for the pristine film. A cross section of a 2D hexagonal mesoporous film is illustrated at bottom inset of figure 2a, showing the hexagonal organisation of the cylindrical pores and the parameters T and d. It should be noted that the parameter d is not equal to lattice parameter of the 2D hexagonal structure, but to the distance between two layers of pores in the direction perpendicular to the film. After irradiation with He ions (figure 2a), it appears that the two first Bragg reflections remain visible, showing that the mesopores are still present and organized in the z direction. The decrease of their intensity and their broadening indicate however that the organisation of the pores is altered. Their shift to the higher Q values is due to a collapse of the periodicity d of the stacking of pores. After irradiation with Ar ions for fluence higher than 10^{14} $Ar.cm^{-2}$, the mesoporous structure is completely destroyed (figure 2b).

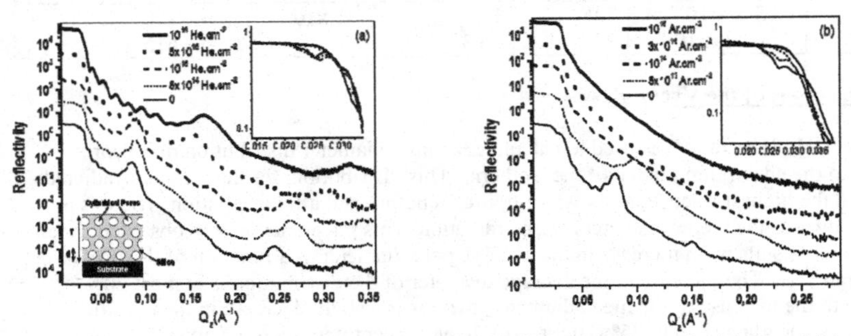

Figure 2. XRR curves of 2D hexagonal mesoporous thin film versus fluence for (a) He and (b) Ar ions.

The decrease of the periodicity d as a function of the fluence is shown on figure 3c. The total thickness T of the film is also decreased from T=89 ± 6 nm to T=49 ± 6 nm (figure 3a). The film is therefore made up with 12 layers of pores before and after irradiation. Another important feature can be observed at small angles (top insets of figure 2a) where two different critical angles of total external reflection Qc are observed. The first one (Qc1) can be related to the average electron density of the film, whereas the second one (Qc2) is assigned to the silicon substrate (~0.032 $Å^{-1}$) [7]. A comparison of the panels, clearly shows that Qc1 is strongly increased after irradiations. Being related to this critical angle by the equation 1, the average electron density of the film is plotted as a function of the fluence in figure 3b. Taking into account the electron density of non-porous silica ρ_{SiO_2} =0.66 $e^-/Å^3$, the film porosity θ [7] can be express by equation 2. The increase of electron density of the film can be related to a porosity decrease given in the table II.

$$\rho_{film} = \frac{1}{16\pi r_e} Qc^2 \text{, where } r_e=2.81.10^{-15}\text{m} \quad (1)$$

$$\theta = \frac{\rho_{SiO_2} - \rho_{film}}{\rho_{SiO_2}} \quad (2)$$

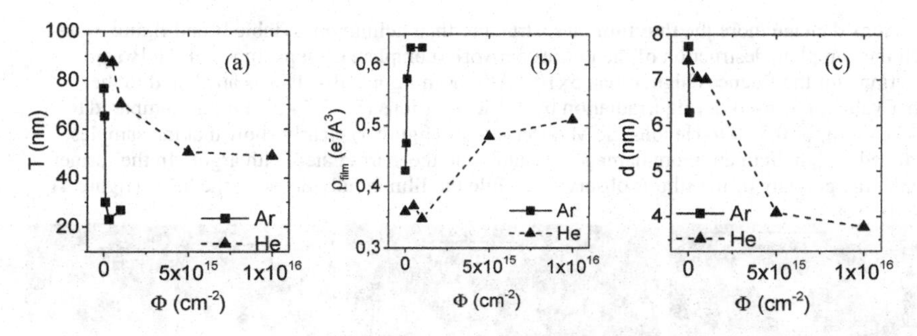

Figure 3. Characteristics of the film versus fluence, calculated from XRR curves for He and Ar irradiations, a) Thickness (T), b) Electronic density (ρ_{film}), c) Pores distances (d).

Table II. Critical angles Q_{c1}, electron density ρ_{film} and deduced porosity θ of the films after He and Ar irradiations under increasing fluences **Φ**.

Ions	**He**					**Ar**				
Φ (cm^{-2})	0	$5x10^{14}$	10^{15}	$5x10^{15}$	10^{16}	0	$5x10^{13}$	10^{14}	$3x10^{14}$	10^{15}
Q_{c1}	0.0225	0.0227	0.0221	0.0260	0.0268	0.0245	0.0257	0.0285	0.0297	0.0279
ρ_{film}	0.359	0.368	0.347	0.482	0.509	0.427	0.471	0.576	0.628	0.628
θ	0.45	0.44	0.47	0.27	0.23	0.35	0.29	0.13	0.05	0.05

DISCUSSION

Contrary to Klaumunzer results, the mesostructure of Vycor glasses irradiated in this study is only slightly modified. This difference can be a consequence of the lower ion energy used; 548MeV-Ar ions leads to a value of S_e=1.5keV/nm, which is lower than the threshold value for the formation of track in silica. The slight increase in the average pore diameter and the small decrease in the specific surface area are consistent with each other, and can be explained by a change in surface roughness of the Vycor glasses. Indeed, Vycor glasses are prepared by leaching of a soluble phase obtained after a phase separation in a borosilicate glass. This technique provides a mesoporous silica with a controlled pore size. However, the surface of the glass constituting the porous network is very rough and often exhibits nanoscale glass nodules resulting from the incomplete dissolution of the labile phase. It is likely that irradiation smooth the surface which generates the observed changes.

XRR permits to monitor the evolution of the total thickness, porosity and porous structure of mesoporous thin films, induced by irradiation. For He irradiations, it appears that these three parameters evolve with the same tendency: up to the fluence 5.10^{15} cm^{-2}, He irradiation provokes a strong decrease of the film thickness associated to a collapse of the pores. Above this fluence, the radiation damage seems to reach a plateau. For Ar irradiations, the same behaviour is observed but occurring at lower fluences. In the irradiations conditions tested, irradiation with argon ions damage more the thin film mesostructure than helium ones (table II and figure 3) and reach the complete destruction of the porous network (complete disappearance of the Bragg reflections for the fluences higher than $5x10^{13}$ Ar/cm² in figure IIb). This is attributed to the higher value of S_e used during irradiation by 20MeV-Ar ions (S_e=3.4 keV/nm) as compared to 0.26 keV/nm for 0.5MeV-He ions. SEM observations (figure 4) clearly show that the samples irradiated with helium are more damage resistant than those irradiated with argon. In the former sample, the porosity of the film is observable while the film is fully dense in the latter (figure 4).

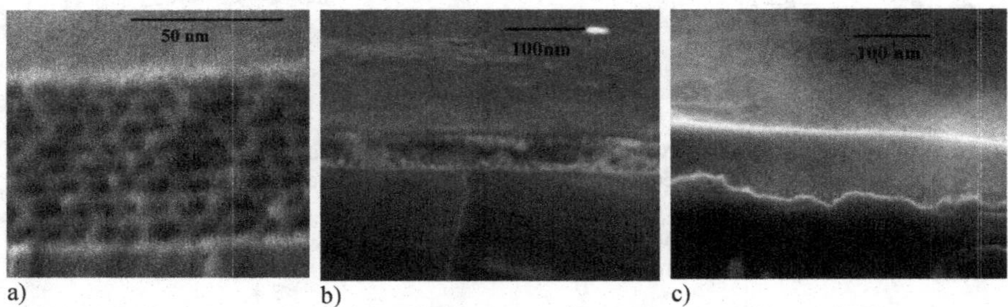

a) b) c)

Figure 4. SEM observations of mesoporous thin film a) Virgin sample; b) Sample irradiated by 0.5MeV-He, $10^{16}cm^{-2}$, c) Sample irradiated 20MeV-Ar, $10^{15}cm^{-2}$.

The results show that the films obtained by the sol-gel are more sensitive to radiation damage than the Vycor glasses. For similar initial characteristics on the virgin mesoporous structures (pore size and pore volume), more significant changes were observed on the meso-structure of the thin films after irradiation in comparison to Vycor glasses. This result was observed after bombardment with particles of high and low electronic stopping power, i.e. respectively Ar and He irradiations. This difference can be explained by the higher rigidity of the network of the Vycor glasses that are elaborated at high temperature. Structural analysis is in progress on these two types of materials to assess changes in the degree of crosslinking of the silica network during the irradiation, and relate this parameter to the rigidity of the network.

REFERENCES

1. J. Roberto, T Diaz de la Rubia, DOE report, “Basic research needs for advanced nuclear energy systems” (2006).
2. P. Makowski, X. Deschanels and al. New J. Chem. 36-3, 531 (2012).
3. S. Klaumunzer, Nucl. Instrum. Methods Phys. Res. B 191, 356 (2002).
4. S. Klaumunzer, Nucl. Instrum. Methods Phys. Res. B 225, 136 (2004).
5. S.O. Kucheyev, Y.M. Wang and al., J. Phys. D:Appl. Phys. 44, 085406 (2011).
6. S. Dourdain, X. Deschanels and al., J.N.M. 427, 411 (2012).
7. S. Dourdain, A. Mehdi and al., Thin Solid Films 495, 205 (2006).
8. E. Barrett, L. Joyner and P. Halenda, J. Am. Chem. Soc. 73-1, 373 (1951).
9. S. Fall, M. Kulijand and A. Gibaud, J. Phys. Condens. Matter 22, 474005 (2010).

Mater. Res. Soc. Symp. Proc. Vol. 1514 © 2013 Materials Research Society
DOI: 10.1557/opl.2013.357

Surface Sensitive Spectroscopy Study of Ion Beam Irradiation Induced Structural Modifications in Borosilicate Glasses

Amy S. Gandy[1] Martin C. Stennett[1] and Neil C. Hyatt[1]
[1]Immobilisation Science Laboratory, Department of Materials Science and Engineering,
The University of Sheffield,
Mappin Street,
Sheffield S1 3JD, UK

ABSTRACT

Fe K edge X-ray absorption (XAS) and Fourier Transform Infra-Red (FT-IR) spectroscopies have been used to study potential structural modifications in sodium borosilicate glasses as a consequence of Kr^+ irradiation. Glasses were doped with simulant waste elements and irradiated at room temperature with 450 keV Kr^+ ions to a fluence of $2x10^{15}$ Kr^+ ions cm^{-1}. According to SRIM calculations, a damaged surface region approximately 400nm wide was produced. In order to probe only the damaged surface layer, XAS measurements were taken in total electron yield mode and FT-IR spectroscopy was conducted in reflectance off the glass surface. No change in Fe valence state was detected by XAS following irradiation. Reflectance FT-IR data revealed a shift to higher wavenumbers in the absorption bands located between 850 and 1100 cm^{-1} in the doped glasses, corresponding to bond stretching in the silicate network. Deconvolution of FT-IR spectra revealed the shift was due to polymerisation of the silicate network. Network connectivity was found to decrease in the un-doped glass, following irradiation. The results suggest an increase in silicate network connectivity by a cation mediated process, and demonstrates the successful application of surface sensitive XAS and FT-IR to the investigation of ion beam induced damage in amorphous materials.

INTRODUCTION

In the UK, alkali borosilicate glasses are used to vitrify high level waste (HLW) produced by reprocessing spent nuclear fuel. HLW contains fission products and minor actinides which continue to undergo radioactive decay in the wasteform for up to 10^6 years. Cations such as Fe and Zr are also present in the waste stream and require incorporation into the final wasteform. Actinides undergo α-decay with the formation of α-particles (He nuclei) and energetic (~100 keV) daughter recoil nuclei. Energy is transferred from the energetic recoil nuclei to other atoms in the glass via elastic interactions, resulting in atomic displacements which form collision cascades. It is hypothesized that accumulation of this ballistic damage can lead to migration of alkali ions, resulting in changes in glass network polymerisation. These changes can affect wasteform durability and since the wasteform acts as the final barrier against radionuclide release into the environment, it is important that the effects of α-decay on the structure of the glass are understood.

Radiation damage in crystalline materials has been extensively studied using a range of complimentary techniques, including ion beam irradiation [1]; actinide doping of materials [2]; investigation of natural analogues [3]; and computational modeling e.g. MD simulations [1]. As a consequence, crystalline material response to fluence and flux is well understood, including the transformation from a crystalline to amorphous structure at sufficiently high damage levels. However, investigation of radiation damage in amorphous materials is more complex.

Heavy ion implantation (e.g. Kr^+ irradiation) provides an analogue for α-recoil damage [4]. Kr^+ ion implantation produces a thin damaged surface region, typically of the order of 1µm. Characterisation of this layer requires application of surface sensitive techniques. XAS allows examination of amorphous materials and provides information about element speciation (e.g. co-ordination number, oxidation state and bond length). In borosilicate glasses, iron can exist as Fe^{3+} as a network former and Fe^{2+} as a network modifier [5]. Therefore, a change in iron oxidation state following irradiation may be indicative of structural modifications. In this study, Fe oxidation state was determined by near edge X-ray absorption spectroscopy (XANES). XANES data was collected in total electron yield mode, where the probe depth corresponds to the Auger electron escape depth, calculated in the systems studied here to be 190nm. FT-IR has been widely used to investigate the structure of borosilicate glass [6 - 8]. In particular, connectivity of the silicate network can be inferred by careful analysis of absorption profiles [7, 8]. Due to the high absorption by oxide glasses in IR, such as those in this study, reflectance FT-IR can be used to probe up to a depth of approximately 100nm [9]. Existing glass wasteforms are multi-component systems with complex atomic structures. In this study, the effect of Kr^+ irradiation on the structure of simple sodium borosilicate glasses has been examined using complimentary surface sensitive spectroscopic techniques, in order to facilitate an understanding of the effects of α-decay in more complex systems.

EXPERIMENT

Three sodium borosilicate glasses were produced according to the compositions given in Table I. DB8 and DB11 were produced using DB3 as the base glass composition and including 2 mol% and 16 mol% simulant waste elements, respectively. Glasses were melted at 1250 °C for 5 hrs (1 hour static and 4 hours stirred to ensure homogeneity) in re-crystallised alumina crucibles using an electrical muffle furnace. Glasses were cast into a steel mould and annealed in a muffle furnace at 500 °C for 1 hr. The annealing temperature was chosen to be close to the glass transition temperature (Tg) for each of the compositions.

Table I. Glass compositions (mol%)

Glass	SiO_2	Na_2O	B_2O_3	ZrO_2	MoO_3	CeO_2	Fe_2O_3	ZnO
DB3	73.65	16.46	9.89	-	-	-	-	-
DB8	72.18	16.13	9.69	0.50	0.125	0.125	0.625	0.625
DB11	61.87	13.83	8.31	4.00	1.00	1.00	5.00	5.00

Samples were cut into disks approximately 1 cm in diameter and 2 mm thick, and polished to a 1 µm finish. Samples were irradiated at room temperature with 450 keV Kr+ ions to a fluence of $2x10^{15}$ Kr^+ ions cm^{-1}. Irradiations were carried out at the Ion Beam Centre at Helmholtz-Zentrum Dresden-Rossendorf, Germany. Nuclear waste glasses are expected to be subjected to 0.1 – 1 dpa (displacements per atom) during disposal lifetime. Results from calculations with the Monte Carlo code TRIM [10] indicated that relevant levels of damage will be produced in the glasses in a region extending from the surface to a depth of approximately 320 nm.

Fe K edge XAS, taken in total electron yield mode, were acquired on beam line 11.1R at ELLETRA synchrotron light source. Data were also acquired in transmission mode using finely ground glass samples dispersed in Polyethylene glycol (PEG) to achieve a thickness of one absorption length. For comparison, spectra were also collected in transmission from a crystalline mineral standard $NaFeSi_2O_6$ (aegerine). XAS data analysis was performed using the programs Athena, Artemis and Hephaestus [11]. FT-IR spectroscopy was used to examine the silicate network. A Perkin Elmer Spectrum 2000 FT-IR spectrometer was used with spectra being collected over the wavenumber range 500–1500 cm^{-1}. To examine the surface regions of both pristine and irradiated samples, experiments were performed by reflectance off the glass surface. Reflectance FT-IR spectra exhibit a shift in band location towards higher wavenumbers compared to transmission FT-IR spectra [12]. In this study, results were corrected using a Kramers-Kronig transformation, to shift absorption bands to positions comparable to those observed in transmission spectra. Examination of the silicate network was achieved by careful deconvolution of the absorption profiles using the 'Peak Fit' program. Gaussian curves have previously been found to provide the best fit when deconvoluting experimental data from glasses [13]. In this study, the position, width and intensity of Gaussian curves, assigned to well established absorption bands reported in the literature, were parameterised manually. The program was allowed to refine these initial values, within acceptable pre-defined limits, in order to achieve the best fit to the experimental data.

RESULTS AND DISCUSSION

Fe K edge XANES

Figure 1 shows a comparison of Fe K edge a) XANES region and b) pre-edge spectra, taken in total electron yield mode for DB8a (pristine); DB8b (irradiated); and taken in transmission mode for DB8a and aegerine. Similar spectra were taken for DB11a and DB11b. Aegerine contains Fe^{3+} in six-fold co-ordination. All spectra have an absorption edge E_0 at 7123.8 eV characteristic of Fe^{3+} species, calculated by taking the first derivative of the spectrum. This suggests that either there is no change in oxidation state from Fe^{3+} due to irradiation, or any modification is below detection limits of this technique.

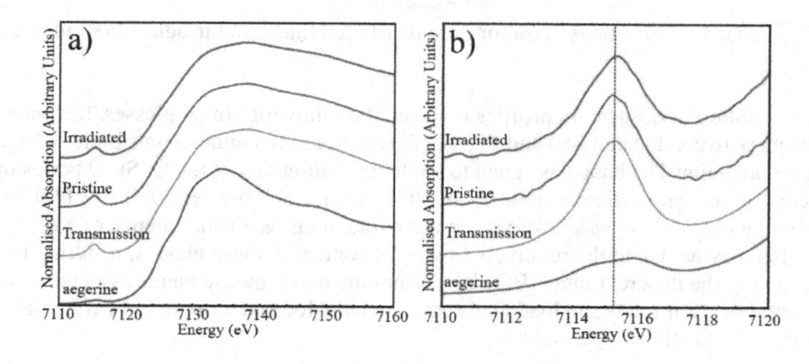

Figure 1. XAS spectra collected in total electron yield mode for pristine and irradiated samples and in transmission mode for pristine samples and aegerine: a) DB8 XANES region and b) DB8 pre-edge spectra.

Reflectance FT-IR

Normalised FT-IR spectra for pristine (dotted lines) and irradiated (solid lines) glasses are presented for all glass compositions (DB3, DB8 and DB11) in Figure 2. One main absorption band is observed between 850 and 1240 cm^{-1}. Two shoulders are observed in all spectra, located between 770 and 790 cm^{-1} and between 1160 and 1250 cm^{-1}. In the two waste loaded glasses, DB8 and DB11, a shift of the main absorption band to higher wavenumbers is observed following irradiation, whereas no shift in peak position is detected in the un-doped glass, DB3. Silicate glass structure can be described in terms of Qn, where n equals the number of bridging oxygen's (BO's) in the SiO_4 tetrahedra. Deconvolution of the absorption bands allowed examination of the silicate network. The area under each Gaussian curve is proportional to the concentration of associated bond species [7]. However, the area is not a direct relation due to overlapping of bands associated with different bonds. In addition, different species bonded to BO's will result in variations in band position. Although FT-IR cannot be used, on its own, to determine quantitatively concentrations of specific bonds, it can be used to infer changes to silicate network connectivity.

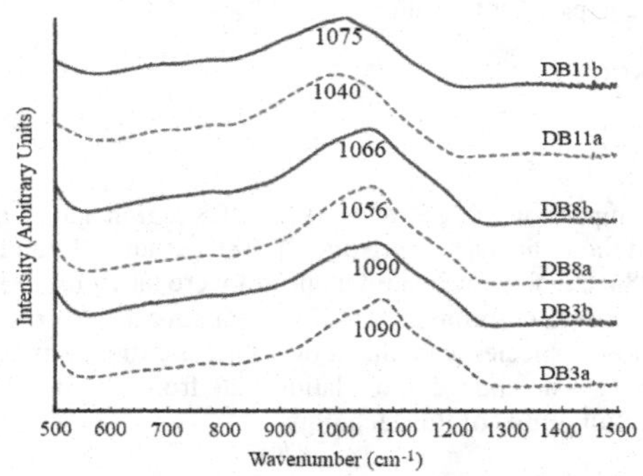

Figure 2. Normalised absorption spectra for pristine (dotted lines) and irradiated (solid lines) glasses.

Deconvolution of absorption profiles revealed the following: In all glasses, the main absorption band between located 850 and 1250 cm^{-1} was found to contain contributions from silicate and borate units; The bands assigned to stretching vibrations of the O–Si–O bonds in SiO_4 tetrahedral units, with different numbers of BO's, are located between 900 and 1150cm^{-1}; In all cases, minimal or no effect was observed on the borate units as a consequence of Kr^+ irradiation. This may be due to the relatively low B_2O_3 content in these glasses, resulting in difficulty resolving the different units (BO_3/BO_4) from the more intense bands associated with the silicate networks; For the waste loaded glasses, the band located around 750 cm^{-1} was assigned to the Fe^{3+}-O stretching bond [14].

In order to investigate the effect of Kr^+ irradiation on silicate network connectivity, Qn species were defined using the following band assignments: 900 to 950cm^{-1} for Q_0; 950 to 1000cm^{-1} for Q_1; 1000 to 1050cm^{-1} for Q_2; 1050 to 1100cm^{-1} for Q_3; and 1100 to 1150cm^{-1} for Q_4 [15]. By comparing the area under the curves assigned to these bands, potential changes in connectivity were inferred. In the un-doped glass a decrease in Q_3 and Q_4 species was observed with an increase in Q_0 to Q_2 species, following irradiation. Conversely, for the doped glasses an increase in Q_3 and Q_4 species was revealed concurrently with a decrease in Q_2 species, resulting in the shift of the main absorption band to higher wavenumbers following irradiation, shown in Figure 2. These results suggest a decrease in silicate network connectivity following irradiation in the un-doped glass, and polymerisation of the silicate network in the doped glasses. Increase in silicate network connectivity can be explained by the migration of alkali cations during irradiation, facilitating re-connection of non-bridging oxygen's (NBO's) into the network. The fact that polymerisation is only observed in doped glasses suggests this process may be mediated by metal cations, such as Fe, Zr or Zn, present in the glass.

CONCLUSIONS

Fe K edge X-ray absorption spectroscopy, in total electron yield mode, and reflectance Fourier Transform Infra-Red spectroscopy were used to probe the surface regions of doped and un-doped sodium borosilicate glasses, in order to examine potential structural modifications as a consequence of Kr^+ irradiation, used to simulate α-daughter recoil nuclei. No variation in Fe oxidation state was detected, potentially due to any variation being below detection limits of this technique. The un-doped glass exhibited a decrease in silicate network connectivity following irradiation, whilst the doped glasses displayed an increase in silicate network polymerisation. Changes to network connectivity can be explained in terms of alkali cation migration during irradiation, mediated by the presence of metal cations within the doped glass. This work demonstrates successful application of surface sensitive spectroscopic techniques to the study of radiation damage in amorphous materials.

ACKNOWLEDGMENTS

This work was supported in part by EPSRC under grant EP/I012214/1. NCH is grateful to The Royal Academy of Engineering and Nuclear Decommissioning Authority for support. This work has also been supported by the European Community as an Integrating Activity 'Support of Public and Industrial Research Using Ion Beam Technology (SPIRIT)' under EC contract no. 227012

REFERENCES

1. W. J. Weber, R. C. Ewing, C. R. A. Catlow, T. Diaz de la Rubia, L. W. Hobbs, C. Kinoshita, Hj. Matzke, A. T. Motta, M. Nastasi, E. K. H. Salje, E. R. Vance, S. J. Zinkle, J. Mater. Res, 13, 1434 (1998)
2. K. Kuramoto, H. Mitamura, T. Banba and S. Muraoka, Progress in Nuclear Energy, 32, 509 (1998)
3. B. I. Omel'yanenko, T. S. Livshits, S. V. Yudintsev, and B. S. Nikonov, Geology of Ore Deposits, 49, 173 (2007)

4. D.P. Reid, M.C. Stennett, B. Ravel, J.C. Woicik, N. Peng E.R. Maddrell, N.C. Hyatt Nucl. Instr. Meth. Phys. Res. B **268**, 1847 (2010)
5. F. Farges, S. Rossano, Y. Lefr`ere, M. Wilke and G. E. Brown Jr, Physica Scripta, T115, 957, (2005)
6. H. Darwish, M. M. Gomaa, J. of Mater. Sci.: materials in Electronics, 17, 35 (2006)
7. K. El-Egili, Physica B 325 340 (2003)
8. M.S. Gaafara and S.Y. Marzouk, Physica B 388, 294 (2007)
9. S. A. MacDonald, C. R. Schardt, D. J. Masiello and J. H. Simmons, Journal of Non-Crystalline Solids, 275, 72 (2000)
10. J.F. Ziegler, J.P. Biersack, U. Littmark, The Stopping and Ranges of Ions in Solids, Pergamon Press, New York, 1985. Available from: <http://www.srim.org>.
11. B. Ravel and M. Newville, Journal of Synchrotron Radiation, 12, 537 (2005)
12. M. Mastalerz. "*Application of reflectance micro-Fourier Transform infrared (FTIR) analysis to the study of coal macerals*" 53rd ICCP Meeting, Copenhagen, August (2001).
13. Y.M. Moustafa, A.K. Hassan, G. E1-Damrawi, and N.G. Yevtushenko, J. of Non-Cryst. Solids, 194, 34 (1996)
14. F.H. El-Batal, E.M. Khalil, Y.M. Hamdy, H.M. Zidan, H.S. Aziz, A.M. Abdelghany, Silicon, 2, 41 (2010)
15. A. A. Akatov, B. S. Nikonov, B. I. Omel'yanenko, S. V. Stefanovsky, and J. C. Marra, Glass Physics and Chemistry, 35, 245 (2009)

Mater. Res. Soc. Symp. Proc. Vol. 1514 © 2013 Materials Research Society
DOI: 10.1557/opl.2013.127

Titanium and zirconium oxidation under argon irradiation in the low MeV range

Dominique Gorse-Pomonti[1], Ngoc-Long Do[1], Nicolas Bérerd[2], Nathalie Moncoffre[2] and Gianguido Baldinozzi[3]
[1]Laboratorie des Solides Irradiés, UMR CNRS 7642, Ecole Polytechnique, F-91128, Palaiseau Cedex, France.
[2]Institut de Physique Nucléaire de Lyon, UMR CNRS 5822, F-69622, Villeurbanne Cedex, France.
[3]Matériaux Fonctionnels pour l'Energie, SPMS CNRS-Ecole Centrale Paris, Châtenay-Malabry, France

ABSTRACT

We studied the irradiation effects on Ti and Zr surfaces in slightly oxidizing environment (rarefied dry air, 500°C) using multi-charged argon ions in the low MeV range (1 – 9 MeV) to the aim of determining the respective role of the electronic and nuclear stopping power in the operating oxidation process under irradiation. We have shown that ballistic collisions contribute significantly to the enhanced Ti and Zr oxidation under MeV argon bombardment. We have also shown that the projectile energy plays a significant role in the overall process.

A significant oxide film thickening is visible on titanium under irradiation, taking the form of a well-defined oxidation peak between 1 and 4 MeV, as a result of the Nuclear Backscattering Spectroscopy and Spectroscopic Ellipsometry studies.

A significant oxide film thickening is also visible on zirconium under same irradiation conditions, at 4 and 9 MeV, as a result of the NBS study. Work is in progress in order to determine how the modified oxidation process depends in this case on the projectile energy.

INTRODUCTION

In a first paper, we proved that large rather circular craters (of diameter in between 200 and 400 nm and of depth not exceeding a few tens of nm) form at the surface of titanium bombarded by an argon ion beam (3×10^{10} ions $cm^{-2}s^{-1}$) of 4 MeV or 9 MeV during 3 hours at 500°C under rarefied air pressure [1]. This result was rather unexpected and encouraged us to continue this study of the radiation damages caused to metallic surfaces by multi-charged heavy ions in the low MeV range under similar environmental conditions. To our knowledge, superficial damages in this energy range were until now very little studied. Indeed, as seen in figure 1, if we plot the nuclear (S_n) and electronic (S_e) stopping power versus ion energy for the Ar/Ti system already studied, and for the Ar/Zr system studied here, we observe that the radiation damages should be produced as a result of combined nuclear and electronic energy losses between 1 and 9 MeV [2]. Moreover those damages could possibly differ largely from the cases in which only one stopping mechanism is operating.

In this short paper, we will show that argon ion irradiation in the low MeV range produces damages on both titanium and zirconium surfaces under rarefied air, that take the form of accelerated oxidation, and more specifically will appear unexpectedly as an oxidation peak as function of the argon ion energy around 3 MeV on titanium but also very presumably on zirconium, under the above mentioned environmental conditions.

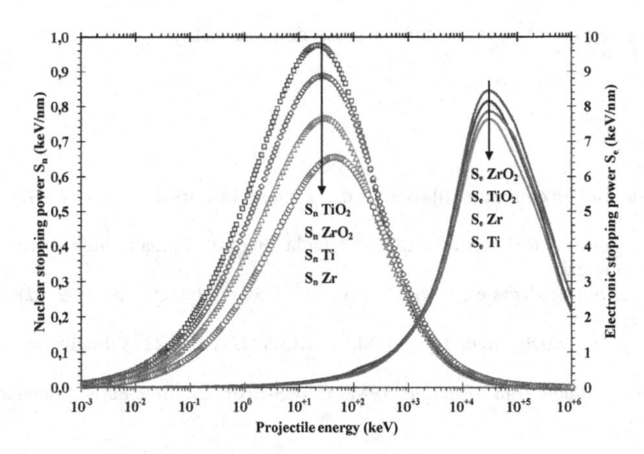

Figure 1 S_n and S_e versus ion energy for different systems: Ar/Ti, Ar/TiO_2, Ar/Zr and Ar/ZrO_2, estimated using SRIM-2011.

EXPERIMENT

Bulk titanium and zirconium plates of 10 x 5 x 1 mm^3 (Ti > 99.6% purity, Zr > 99.8 purity, Goodfellow Corp.) were mechanically polished to mirror. Thin titanium foils of 2.5 μm thickness were also gently cleaned. These specimens were oxidized under argon irradiation at the Nuclear Physics Institute of Lyon (IPNL), using the irradiation chamber adapted to the extracted beam line of the 4 MV Van de Graff accelerator of IPNL, allowing to monitor the specimen temperature and gas pressure in the irradiation chamber. The irradiation conditions are: 1 to 9 MeV Ar^{n+} (n=1, 3) at a dose rate of 3 x 10^{10} ions cm^{-2} s^{-1} up to a dose of 5 x 10^{14} ions cm^{-2}, under controlled rarefied dry air pressure of 5 x 10^{-3} Pa at 500°C.

The oxygen gain of both Ti and Zr specimens was measured by Nuclear Backscattering Spectroscopy (NBS) using 7.5 MeV alpha particles, taking advantage of the high oxygen cross section of the nuclear reaction $^{16}O(\alpha, \alpha')$ at this energy, compared to the classical Rutherford backscattering cross section [3].

The Ti specimens were analysed by Spectroscopic Ellipsometry (SE). A commercial UVISEL ellipsometer (HORIBA, Jobin-Yvon) was used for measurements in between 400 and 800 nm (1.5 to 3 eV). The experimental spectra were analyzed using the Horiba DeltaPsi 2 commercial software. The SE spectra obtained with the specimens oxidized under argon irradiation at 9 and 4 MeV exhibit unexpectedly an absorbing metallic character. In order to determine their oxide thickness, we have chosen a model composed of one compact TiO_{2-x} layer the dielectric function of which is represented by the combination of the Drude [4] and Forouhi-Bloomer [5] oscillators on the real Ti substrate whose dielectric function was determined previously. The same model was used for the reference Ti specimen oxidized under same conditions without irradiation (NOT in Table I). The SE spectra obtained with the specimens oxidized under argon irradiation at 1, 2 and 3 MeV contrast with the ones obtained at higher energy: they exhibit a pronounced transparent dielectric character, and were modelled using an external compact TiO_x layer on an intermediate TiO_{2-x} layer on a real Ti substrate as above, the dielectric function of the external layer being represented by a Forouhi-Bloomer oscillator and that of the internal layer being represented by a combination of one Drude and one Lorentz oscillator. The χ^2 values reported in table I allow for estimating the fit quality. The detailed analysis of the SE spectra will be the subject of a shortcoming publication [6].

DISCUSSION

We report chronologically the results of recent titanium and zirconium oxidation under argon irradiation studies showing that the nuclear stopping S_n *and* the projectile energy play a role in the corrosion under irradiation of Ti and Zr in the low MeV range, in environmental conditions reproducing the fuel/cladding interfaces of PWR fuel rods.

Zirconium oxidation under argon ion irradiation: S_n effect

By using Nuclear Backscattering Spectroscopy (NBS), it was first proved in 2006 by N. Bérerd *et al.* that collisions cascades generated by 4 or 9 MeV Ar beams in micrometre thick Zr foils, well below the surface (R_p = 1.7 µm or 2.81 µm, respectively), increase significantly the thermal oxidation of zirconium [7]. As seen in figure 2, the oxide thickness is double when the Ar beam stops in the Zr foil, by comparison with the case of the Ar beam passing through the foil. The environmental (5×10^{-3} Pa of dry air, 500°C) and irradiation (3×10^{10} ions cm^{-2} s^{-1} up to a dose of $5\ 10^{14}$ ions cm^{-2}) conditions were the same as used in the present work.

To our knowledge this was the first proof of an S_n effect on oxide growth, and particularly surprising in this energy range for which the target surface was considered a priori transparent to the Argon beam. This result was confirmed when the Ti specimens were irradiated by Ar beams of energy varying in between 1 and 9 MeV, under otherwise similar experimental conditions.

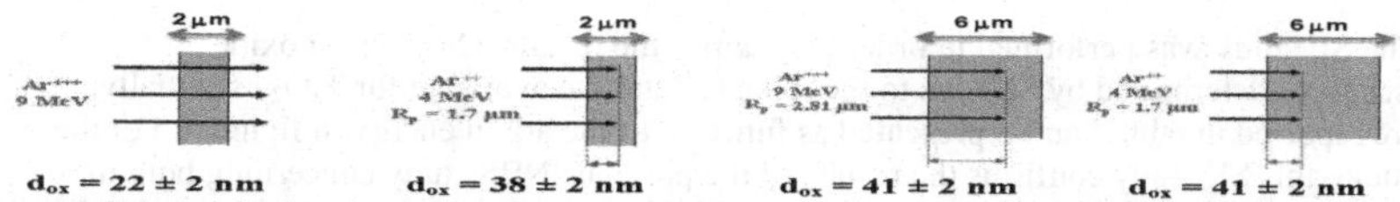

Figure 2 Schematic drawing of the oxidation under argon ion flux of Zr foils showing the effect of the stopping (38 nm $\leq d_{ox} \leq$ 41 nm) or not (d_{ox}= 22 nm) of the Ar beam in the Zr target.

Zirconium and titanium oxidation under argon ion irradiation: Argon energy effect

All Ti and Zr specimens, bulk or thin foils, whatever oxidized (NOT) or oxidized under argon irradiation, were analyzed here by Nuclear Backscattering Spectroscopy, considered as a powerful tool for profiling oxygen.

We obtained the following surprising results represented in figure 3: i) the plot of the oxygen gain measured on Ti specimens as function of the argon ion energy exhibits a well defined peak around 3 MeV; (ii) the oxygen gains decrease to values lower than that of the reference oxidized Ti once reached the energy of 7.5 MeV; (iii) the oxygen gain measured on Zr specimens at 4 and 9 MeV is between 2.5 and 3.4 times larger than the one measured on the reference oxidized Zr specimen without irradiation; (iv) the oxygen gain measured on Zr specimens increases with decreasing argon energy from 9 to 4 MeV, suggesting the existence of an oxidation peak as function of the Ar energy for Zr, as for Ti, in present exp. conditions.

We prove here by NBS that not only collision cascades play a role in oxidation under irradiation, but also that the oxidation acceleration under irradiation is largely and unexpectedly dependent upon the argon beam energy in the 1 - 9 MeV range.

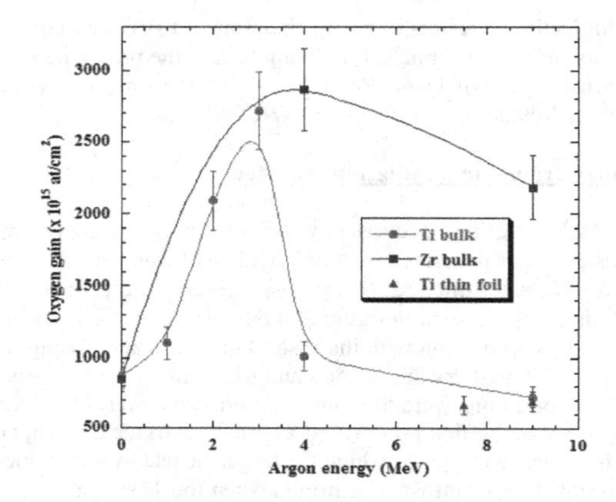

Figure 3 Oxygen gain as function of argon energy for Ti and Zr specimens oxidized under irradiation in between 1 and 9 MeV. The nature of the target, 1 mm thick Ti and Zr (bulk) or 2.5 μm thick Ti (foil) is noted. The oxygen gains of the reference oxidized Ti, Zr are noted at E = 0.

The SE study was performed in order to quantify the results. Only the Ti oxide thicknesses were determined by SE, due to the reliable database available for TiO_2 essentially. They were reported in table I and represented as function of the argon energy in figure 4. For the Ar/Ti system, this SE study confirms the results of the previous NBS study concerning both the Sn and argon energy effect. In addition we are now able to prove that the argon bombardment of the Ti surface under rarefied air acts as a powerful oxidizing agent, at least at 3 MeV for which the oxide thickness found equal to 54 ± 1 nm approaches the oxide thickness measured on Ti after 3 hours exposure to a mixture of argon and water vapor (613 Pa) at same temperature [8].

Table I. Titanium oxide thicknesses determined by SE as function of the energy of the Argon ion beam by comparison with that of an un-irradiated specimen (NOT) oxidized under same conditions (3 hours under 5 x 10^{-3} Pa of dry air at 500°C). The χ^2 values are indicated.

E_{Ar} (MeV)	1	2	3	4	9	NOT
d_{ox} (nm)	23 ± 2	26 ± 2	54 ± 2	14 ± 1	10 ± 1	9 ± 1
χ^2	2.92	1.30	0.31	0.25	0.05	0.03

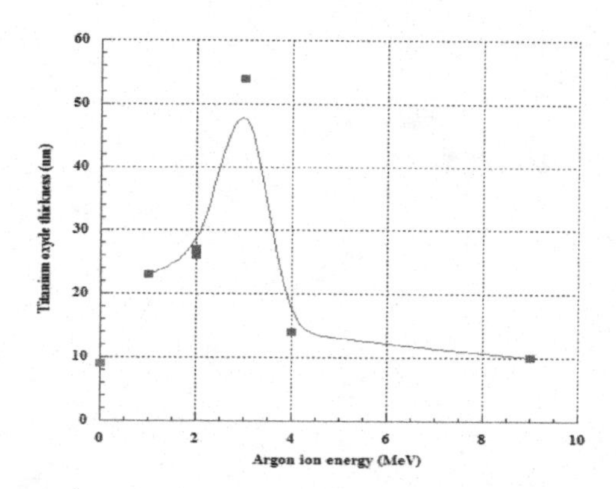

Figure 4 Oxide thickness determined by SE versus Ar energy for Ti specimens oxidized under irradiation (full line: eye guide). The oxide thickness of the ref. oxidized Ti is noted at E = 0.

CONCLUSIONS

Our objective was to determine the respective role of the electronic excitations and ballistic collisions in the oxidation of Ti and Zr under argon irradiation in the low MeV range.

We have shown by coupling two powerful surface techniques, NBS and SE, that ballistic collisions contribute to the modified oxidation process under Ar flux, taking the form of an oxidation peak as function of the argon energy, centered around 3 MeV for the Ar/Ti system.

Work is in progress in order to determine the position of the maximum of the oxidation peak for the Ar/Zr system, under present environmental and irradiation conditions.

REFERENCES

1. N.-L. Do, N. Bererd, N. Moncoffre, F. Yang, P. Trocellier, Y. Serruys, D. Gorse-Pomonti, *J. Nucl. Mater.*, **419**, 168 (2011).
2. J.F. Ziegler, J.P. Biersack, U. Littmark, The Stopping and Ranges of Ions in Solids, Pergamon, New York (1985).
3. A. Chevarier, N. Chevarier, P. Deydier, H. Jaffrézic, N. Moncoffre, M. Stern, J. Tousset, *J. Trace Microprobe Tech.* **6**, 1 (1988).
4. P. Drude, *Ann. Physik u.d. Chem*, **39**, 481 (1890).
5. A.R. Forouhi, I. Bloomer, *Phys. Rev. B*, **34**, 7018 (1986).
6. Ngoc-Long Do, N. Bérerd, N. Moncoffre, E. Garcia-Caurel and D. Gorse, *in preparation for J. Nucl. Mater.*
7. N. Bérerd, N. Moncoffre, A. Chevarier, H. Jaffrézic, H. Faust and E. Balanzat, *Nucl. Instrum. Meth.* **B249**, 513 (2006).
8. F.G. Fuhrman, F.C. Collins, *J. Electrochem. Soc.*, **124**, 1294 (1977).

Mater. Res. Soc. Symp. Proc. Vol. 1514 © 2013 Materials Research Society
DOI: 10.1557/opl.2013.387

Irradiation Induced Effects at Interfaces in a Nanocrystalline Ceria Thin Film on a Si Substrate

Philip D Edmondson,[1,2,§] Neil P Young,[1] Chad M Parish,[2] Fereydoon Namavar,[3] William J Weber,[2,4] and Yanwen Zhang[2,4]
[1]*Department of Materials, University of Oxford, Parks Road, Oxford, OX1 3PH, UK*
[2]*Materials Science and Technology Division, Oak Ridge National Laboratory, Oak Ridge, TN 37831, USA*
[3]*University of Nebraska Medical Center, Omaha, NE 68196, USA*
[4]*Department of Materials Science and Technology, University of Tennessee, Knoxville, TN 37996, USA*
[§]Corresponding author: Philip Edmondson, philip.edmondson@materials.ox.ac.uk

ABSTRACT

Thin films of nanocrystalline ceria on a Si substrate have been irradiated with 3 MeV Au^+ ions to fluences of up to $1x10^{16}$ ions cm^{-2}, at temperatures ranging between 160 to 400 K. During the irradiation, a band of contrast is observed to form at the thin film/substrate interface. Analysis by scanning transmission electron microscopy in conjunction with energy dispersive and electron energy loss spectroscopy techniques revealed that this band of contrast was a cerium silicate amorphous phase, with an approximate Ce:Si:O ratio of 1:1:3.

INTRODUCTION

Ceramics are key materials in the nuclear industry, having applications as core structure materials, inert fuel matrices, potential waste forms for the long term sequestration of high level nuclear waste, novel clad for improved accident tolerance under loss of coolant accident (LOCA) conditions, and as non-radioactive surrogates in the study of defect behavior of fuels under irradiation [1-9]. Whilst the majority of research effort has been devoted to the radiation response of single crystal or relatively large grained polycrystals, there has been little effort in the response of nanocrystalline materials. Nanocrystalline materials are technologically interesting materials due to their ability to control and modify their bulk properties by varying the grain size [10-12] and have been suggested as a viable route towards radiation tolerance [13]. All of these benefits are in part due to the large number of unique grain boundary structures produced in the film [14, 15].

Ceria is typically used as a non-radioactive surrogate in studies of the behavior of defects under irradiation due to a similar behavior when compared with UO_2 and PuO_2 [4, 16, 17]. Ceria is thermodynamically stable in the cubic phase, however the phase stability relies quite heavily on the O/Ce ratio. Ceria retains the cubic phase when CeO_{2-x} is $0<x<0.5$. If $x>0.5$ a phase change to the hexagonal system occurs with the chemical form Ce_2O_3 (there has been reports of a rhombohedral phase forming at $CeO_{1.82}$, however this has not been replicated) [18]. This is commensurate with a change in the oxidation state of the cation from Ce^{4+} to Ce^{3+}.

The work presented in this article describes the response of nanocrystalline ceria to ion irradiation with an emphasis on the oxidation state of the ceria and the thin film/substrate interfacial response.

EXPERIMENTAL METHODS

Thin films of nanostructured ceria (CeO_2) were produced using an ion beam assisted deposition technique at the Nanotechnology Laboratory at the University of Nebraska Medical Center [19]. This resulted in the formation of a ~300 nm thick, nanostructured film grown on top of an approximately 5 nm buffer layer of SiO_2 on a Si substrate. The use of this technique allows for the formation of high purity films, with the control of parameters such as surface morphology, density, stress level, crystallinity, grain size and grain orientation [19]. The use of ion beams also aids in film adhesion to the substrate.

The films were then irradiated with 3 MeV Au^+ ions up to fluences of $1x10^{16}$ ions cm^{-2} at temperatures ranging from 160 – 400 K using the 3 MV Tandem accelerator facilities in the Environmental and Molecular Sciences Laboratory (EMSL) located at Pacific Northwest National Laboratory (PNNL). The ion flux was constant during the irradiation, and the beam was rastered over the surface of the film to ensure a uniform irradiation. The ion beam energy was chosen such that the energy deposited into the film was maximized whilst minimizing Au concentration profiles in the film. The corresponding displacements per atom (dpa) values were calculated using SRIM 2008.01 full cascade simulations [20]. The parameters used were a sample density of ~6.3 gm cm^{-3}, with threshold displacement energies of 27 and 56 eV for the O and Ce respectively. This yielded an average fluence to dpa conversion factor of 0.54 to convert 10^{14} ions cm^{-2} to dpa [17].

Post-irradiation characterization was performed using transmission electron microscopy (TEM), scanning transmission electron microscopy (S/TEM) combined with electron dispersive spectroscopy (EDS) and electron energy loss spectroscopy (EELS). Samples were prepared by mechanically polishing the samples down to a thickness of ~15-20 μm using the tripod polishing technique, followed by ion milling to electron transparency using a Gatan precision ion polishing system (PIPS) operating with a beam energy of 4.5 keV, gradually reduced to 3 keV. The specimens were then evaluated using either a Phillips CM200 FEG S/TEM operating at 200 keV, or a JEOL 3000F FEG S/TEM operating at 300 keV.

RESULTS AND DISCUSSION

A typical cross-sectional TEM micrograph, selected area electron diffraction (SAED) pattern and high resolution image of the as-deposited nanostructured ceria (NSC) film is shown in Figure 1. The micrographs show that the film is nanocrystalline, and has a thickness of approximately 300 nm. The high resolution image shows that the grains were of high quality and in random orientations, also inferred by the SAED pattern. The SAED pattern also indicates that the film is in the cubic phase, and this was confirmed by glancing incidence x-ray diffraction.

The response of the ceria film to irradiation at temperatures of 160 and 400 K is shown in Figure 2. It is evident that the grains have undergone irradiation-induced growth, and this has been discussed elsewhere [16, 17]. No thermally induced grain growth was observed following annealing at 400 K for 5 hours. What can also be observed is that there is a band of contrast that is formed at the interface between the NSC film and the Si substrate in both cases. This band of contrast forms at low doses and increases in thickness with increasing ion dose. It is interesting to note that this occurs with a seemingly different mechanism than the formation of bands of contrast formed under similar conditions in nanocrystalline zirconia films, whereby the bands of contrast form by trapping of the irradiated species (Au^+ ions) at voids, dislocations and interfaces [21, 22].

STEM-EDS and STEM-EELS experiments were then used to characterize the newly

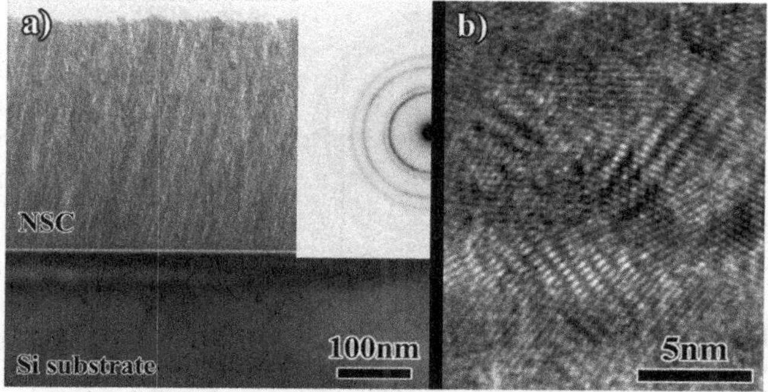

Figure 1: Cross-sectional TEM image of the as-deposited nanostructured ceria (NSC) film on a Si substrate (a) with SAED inset. A high resolution TEM image, (b), shows the grain structure of the NSC film.

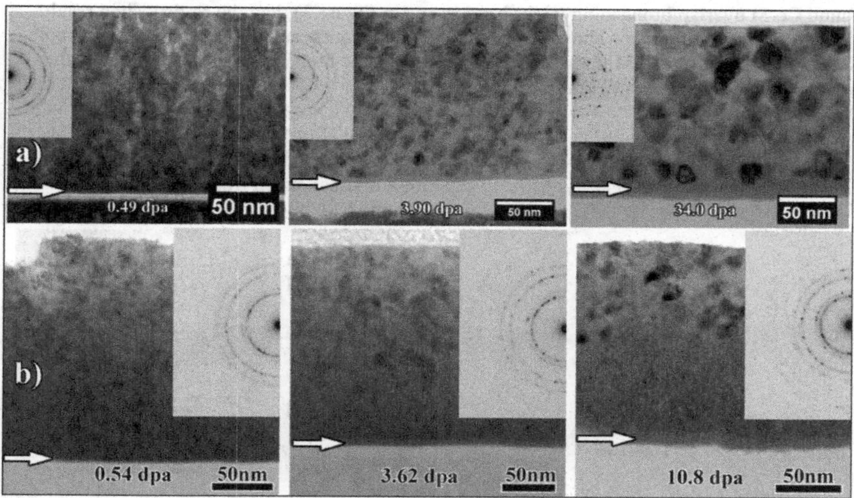

Figure 2: Cross-sectional TEM images of the nanostructured CeO_2 film irradiated at a) 400K, and b) 160K. The doses are indicated in the figure. The arrows indicate the location of the band of contrast formed.

formed phase within the band of contrast, and the results are shown in Figure 3 and Figure 4 respectively.

The STEM-EDS results (Figure 3) show that in the as-deposited case, there is no mixing of the Ce and Si at the interface. There is an overlap in the EDS data that is a result of the spreading of the electron beam through the thickness of the TEM foil. Using the method described by Michael *et al* [23] and assuming a specimen thickness of 100 nm, the minimum spatial resolution

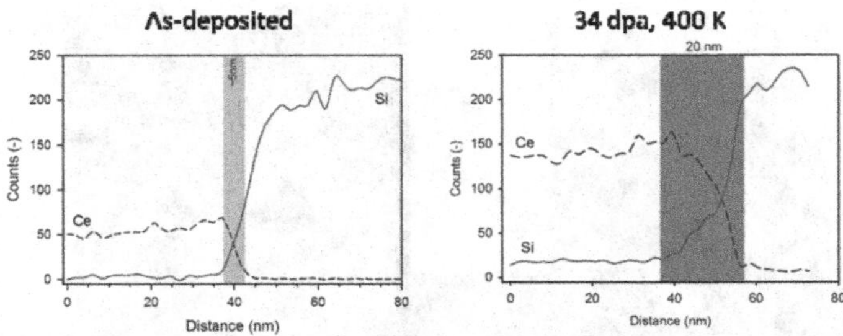

Figure 3: Graphs showing the results of the STEM-EDS data set for a linescan recorded from the NSC film into the Si substrate for the as-deposited materials (left) and the sample irradiated to a dose of 34 dpa at 400 °C (right).

is approximately 5 nm – commensurate with the ghost mixing that is seen in this result. The STEM-EDS results obtained from the 400 K, 34 dpa sample displays a significantly more profound Ce-Si overlap/broadening (right-hand image in Figure 3) than the as-deposited result. The grey box inset of the graph showed the approximate location and thickness of the transition region as measured from TEM images. This suggests that the transition region is a chemically mixed region containing both Ce and Si.

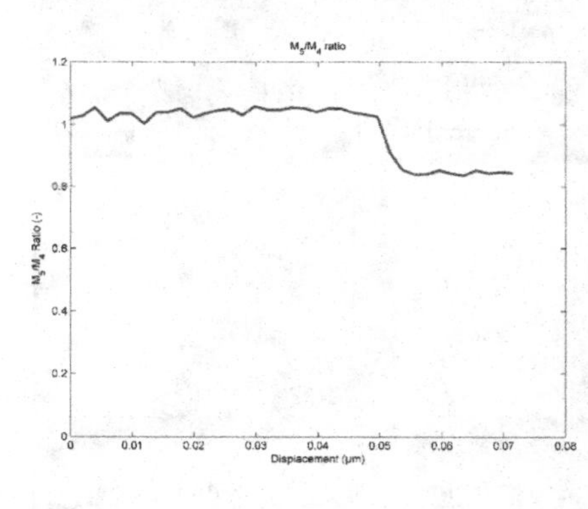

Figure 4: (Left) Graph showing the M_5/M_4 ratio extracted from a STEM-EELS data set for a linescan taken from the Si substrate into the NSC film (from left to right).

To the right is a HAADF STEM image with the spectrum image linescan region from where the graph to the left was recorded indicated.

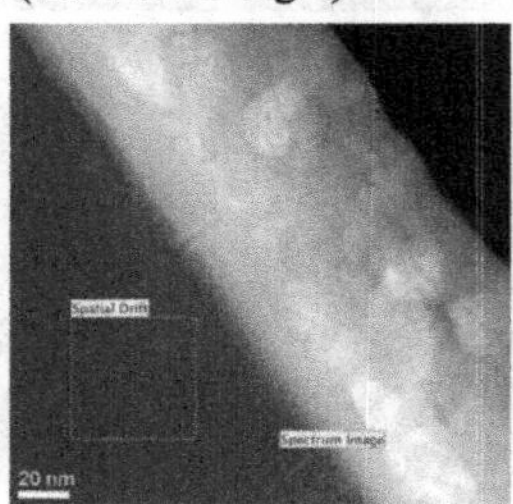

STEM-EELS of the same sample used for the STEM-EDS was also performed, with a focus on the region of the EELS spectrum containing the Ce M_5 and M_4 edges at 885 and 903 eV respectively. The M_5/M_4 ratio is routinely used to determine the oxidation state of ceria [24]. For that reason the M_5/M_4 ratio as a function of scan distance from the NSC film to the Si substrate was calculated using a custom MatLab script that extracts the individual spectra from 2-dimensional spectrum images and calculates the M_5/M_4 ratio, the output of which is shown in Figure 4. Here, a ratio of ~1 is commensurate with Ce^{3+}; 0.8 corresponds to Ce^{4+}.

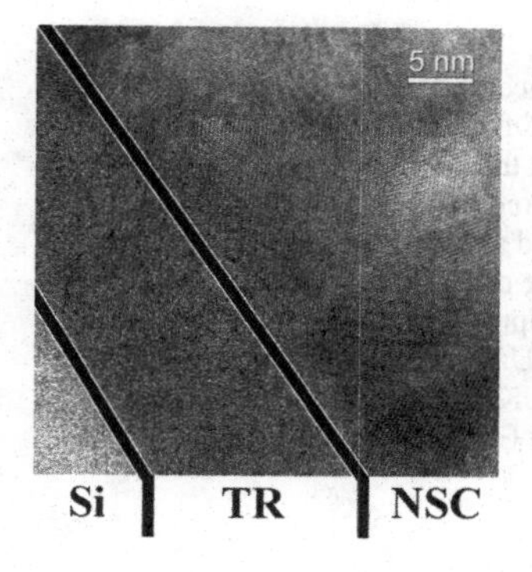

Figure 5: High resolution TEM image of the silicon/thin film interface region. The Si substrate, transition region (TR) and NSC film are labeled. Both the Si and TR are amorphous in nature whilst the NSC remains crystalline.

In the graph shown in Figure 4 for the region 0 to ~0.3μm is sampled from the Si substrate, ~0.3 to ~0.5 μm is sampled from the transition region, and ~0.5 μm onwards is sampled from the NSC film. From this it is clear that the Si substrate and the transition region both contain Ce with the majority being in the 3+ oxidation state. The NSC film remains in the 4+ state. By examination of the oxidation states of the Ce in cerium silicides and cerium silicates, it becomes immediately apparent that it may be one of the following phases: Ce_5Si_3, Ce_3Si_2, Ce_5Si_4, CeSi, Ce_3Si_5, $CeSi_2$, $CeSiO_3$, and Ce_2O_3 [25, 26]. Through integrating the EELS spectra intensity over the energy loss region 960-1000 eVfor background corrected spectra, it is possible to quantify the elemental concentration of the Ce in the transition region. Performing this analysis gives a Ce concentration of 22.0 at.%. This is very similar in concentration to both the Ce_2O_3 and $CeSiO_3$ phases. The EDS results shown in Figure 3 clearly show the presence of Si in the phase, and as such Ce_2O_3 can be eliminated suggesting that the composition of the transition region is consistent with that of the perovskite phase of cerium silicate. High-resolution TEM imaging indicates that the newly formed phase is amorphous in nature, as shown in Figure 5.

This indicates that during irradiation, an ion beam induced chemical mixing of the NSC film/substrate interface occurs, as has been observed in other systems [27]. This may be attributed to the atomic re-ordering of the constituent atoms (Ce from the NSC film, Si from the Si substrate and the SiO_2 layer, and O from the NSC film and SiO_2 layer) in the interfacial region during the ballistic phase of the collision cascade.

SUMMARY

Thin films of nanocrystalline ceria on a silicon substrate have been irradiated with 3 MeV Au+ ions and the irradiation response examined using microscopy techniques. It was revealed that a band of contrast formed at the film/substrate interface, and that this contrast formation was as a direct result in the ion beam induced mixing of the nanocrystalline ceria film and the Si substrate. This amorphous transition region was characterized to have a Ce:Si:O ratio of 1:1:3 – similar to that of the perovskite phase. Analysis of the oxidation state of the Ce in the film and the transition region showed a change from Ce^{4+} in the film, to Ce^{3+} in the transition region, consistent with the formation of a perovskite-type phase.

ACKNOWLEDGMENTS

This work was supported as part of the Materials Science of Actinides, an Energy Frontier Research Center funded by the U.S. Department of Energy, Office of Science, Office of Basic Energy Sciences. A portion of the research was performed at the Environmental Molecular Sciences Laboratory (EMSL), a national user facility sponsored by the Department of Energy's Office of Biological and Environmental Research and located at Pacific Northwest National Laboratory. Part of this research was sponsored by the Office of Basic Energy Sciences, U.S. Department of Energy and by ORNLs Shared Research Equipment (SHaRE) User Facility, which is sponsored by the Office of Basic Energy Sciences, U.S. Department of Energy. PDE and NPY would also like to acknowledge support from the UKs Engineering and Physical Sciences Research Council (EPSRC) for support under grant EP/H018921/1.

REFERENCES

1. Allen, T., *et al.*, MRS Bulletin, 34 (2009) 20-27.
2. Weber, W.J., Journal of Nuclear Materials, 98 (1981) 206-215.
3. Weber, W.J., Journal of Nuclear Materials, 114 (1983) 213-221.
4. Weber, W.J., Radiation Effects, 83 (1984) 145-156.
5. Weber, W.J., Nuclear Instruments and Methods in Physics Research Section B: Beam Interactions with Materials and Atoms, 166-167 (2000) 98-106.
6. Weber, W.J., *et al.*, Journal of Materials Research, 13 (1998) 1434 - 1484.
7. Weber, W.J., *et al.*, MRS Bulletin, 34 (2009) 46-53.
8. Edmondson, P.D., *et al.*, Scripta Materialia, 65 (2011) 675-678.
9. Zhang, Y., *et al.*, Physical Review B, 82 (2010) 184105.
10. Alivisatos, A.P., Science, 271 (1996) 933-937.
11. Knöner, G., *et al.*, Proceedings of the National Academy of Sciences of the United States of America, 100 (2003) 3870-3873.
12. Norris, D.J., A.L. Efros, and S.C. Erwin, Science, 319 (2008) 1776-1779.
13. Bai, X.-M., *et al.*, Science, 327 (2010) 1631-1634.
14. Bai, X.-M., *et al.*, Physical Review B, 85 (2012) 214103.
15. Edmondson, P.D., *et al.*, Acta Materialia, 60 (2012) 5408-5416.
16. Edmondson, P.D., *et al.*, Physical Review B, 85 (2012) 214113.
17. Zhang, Y., *et al.*, Physical Chemistry Chemical Physics, 13 (2011) 11946-11950.
18. Trovarelli, A., Catalysis Reviews, 38 (1996) 439-520.
19. Namavar, F., *et al.*, Nano Letters, 8 (2008) 988-996.
20. Ziegler, J.F., Journal of Applied Physics, 85 (1999) 1249-1272.
21. Edmondson, P.D., *et al.*, Materials Research Society Symposium Proceedings, 1298 (2011) 111-116.
22. Edmondson, P.D., *et al.*, Nuclear Instruments and Methods in Physics Research Section B: Beam Interactions with Materials and Atoms, 269 (2011) 126-132.
23. Michael, J.R., *et al.*, Journal of Microscopy, 160 (1990) 41-53.
24. Haigh, S.J., *et al.*, ChemPhysChem, 12 (2011) 2397-2399.
25. Okamoto, H., Journal of Phase Equilibria and Diffusion, 25 (2004) 98-99.
26. Weitzer, F., *et al.*, Journal of Materials Science, 26 (1991) 2076-2080.
27. Parish, C.M., *et al.*, Journal of Nuclear Materials, 418 (2011) 106-109.

Mater. Res. Soc. Symp. Proc. Vol. 1514 © 2013 Materials Research Society
DOI: 10.1557/opl.2013.199

Atomistic observation of electron irradiation-induced defects in CeO_2

Seiya Takaki, Tomokazu Yamamoto, Masanori Kutsuwada, Kazuhiro Yasuda, Syo Matsumura
Department of Applied Quantum Physics and Nuclear Engineering, Kyushu University,
Fukuoka 819-0395, Japan

ABSTRACT

We have investigated the atomistic structure of radiation-induced defects in CeO_2 formed under 200 keV electron irradiation. Dislocation loops on {111} habit planes are observed, and they grow accompanying strong strain-field. Atomic resolution scanning transmission electron microscopy (STEM) observations with high angle annular dark-field (HAADF) and annular bright-field (ABF) imaging techniques showed that no additional Ce layers are inserted at the position of the dislocation loop, and that strong distortion and expansion is induced around the dislocation loops. These results are discussed that dislocation loops formed under electron irradiation are non-stoichiometric defects consist of oxygen interstitials.

INTRODUCTION

It has been shown that oxide ceramics with fluorite structure, such as yttria-stabilized cubic zirconia (YSZ) and ceria (CeO_2), exhibit exceptional resistance to radiation damage. For example, previous investigations on YSZ have revealed its excellent resistance to amorphization [1] and volumetric swelling [2] under irradiations with energetic particles. Fluorite-type structure oxides, therefore, have potential applications to inert matrix fuels and transmutation targets [3,4].

In fluorite-type oxides, there exists a significant difference in mass between cations and anions. In addition, the displacement energy under energetic particle irradiation has been reported to be larger for cation-sublattice than anion-one [5], leading to the displacement damage rate (dpa/s) for anion sublattice being larger than cation one. A calculation of displacement cross-section in CeO_2 for O- and Ce-sublattice by using McKinly-Feshbach formula [6] revealed that elastic displacements are induced only in O-sublattice with electron energies less than about 1500 keV, whereas both O- and Ce-lattices are displaced with electrons above 1500 keV (Figure 1). Recent studies on electron irradiated CeO_2 reported the formation of non-stoichiometric dislocation loops below 1250 keV, and the defect clusters formed on {111} planes accompanying strong strain-field are discussed to be dislocation loops consist of oxygen ions through transmission electron microscopy (TEM) analysis [7,8]. A molecular dynamic simulation including several oxygen Frenkel pairs has shown that interstitial oxygen ions aggregate to form plate-like clusters on a (111) plane after the relaxation [9], which supports the interpretation based on the TEM analysis.

The present study aims to gain insights into the atomistic structure of radiation-induced defects formed under electron irradiation. Atomic resolution scanning transmission electron microscopy (STEM) with high angle annular dark-field (HAADF) and annular bright-field (ABF) technique was utilized to clarify the atomic structure of dislocation loops. The latter technique has been developed recently to detect the location of light element columns in an atomic scale [10,11].

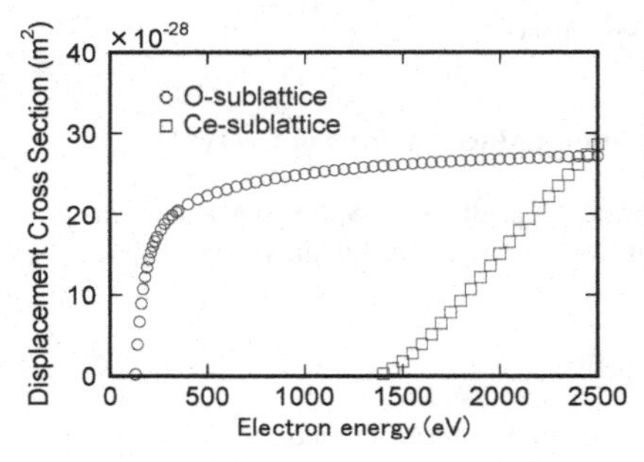

Figure 1. Calculated elastic displacement cross-section for O- and Ce-sublattice in CeO_2 using McKinly-Feshbach formula [6] as a function of electron energy. Displacement enegy is assumed to be 20 and 50 eV for O- and Ce-sublattice, respectively [5,8,12].

EXPERIMENTAL

Powders of CeO_2 with 99.99 % purity (Rare Metal Corp.) were compacted into pellets by uniaxial pressing, and the pellets were subjected to subsequent hydrostatic pressing for densification in a water bath at 150 MPa. The pellets were sintered at 1873 K for 12 hours in air to obtain polycrystalline specimens. The sintered compacts were evaluated to be 98.5 % theoretical density with an average grain size of 5 μm. Disk specimens of 3 mm in diameter and 100 μm in thickness were prepared by mechanical polishing, followed by the dimpling and Ar-ion thinning processes to prepare thin-foil specimens for electron microscopy.

The thin-foil specimens were subjected to "*in situ*" observations under 200 keV electron irradiation at 300 K. Two kinds of electron microscopes in the HVEM Laboratory of Kyushu University, a conventional TEM (JEM-2100HC, JEOL Ltd.) and an atomic resolution TEM with a spherical aberration correction equipment (JEM-ARM200F, JEOL Ltd.), were used in the present study. Nucleation-and-growth process of radiation-induced defects was observed *in situ* under 200 keV electron irradiation with bright-field (BF) imaging technique. HAADF and ABF STEM imaging techniques were also utilized to obtain atomic resolution images of radiation-induced defects formed under electron irradiation. The angles of diffracted electrons to the inner and outer edges of the annular detector for HAADF mode were 90 and 170 mrad, respectively, and 11 and 22 mrad for ABF mode, respectively. The annular detector for ABF mode is a BF detector in conjunction with a circular beam-stop at the beam center.

RESULTS AND DISCUSSION

Nucleation-and-growth process of dislocation loops

Figure 2 shows a sequential change of BF images illustrating the nucleation-and-growth process of dislocation loops formed under 200 keV electron irradiation at an electron beam flux of 1.5×10^{23} electrons/m^2s. It is seen that dislocation loops grow with increasing irradiation time accompanying strong strain field. It is interesting to note that a part of loops moves during irradiation on a $(11\bar{1})$ plane (examples are shown as loop 1 and 2 in Figure 2(b)-(f)). In addition, some dislocation loops, denoted as loop 3 and 4 in Figure 2(e), coalesce during the movements under electron irradiation (Figure 2(f)). The contrast and growth process are different from the perfect dislocation loops with a stoichiometric composition as reported previously [7,8].

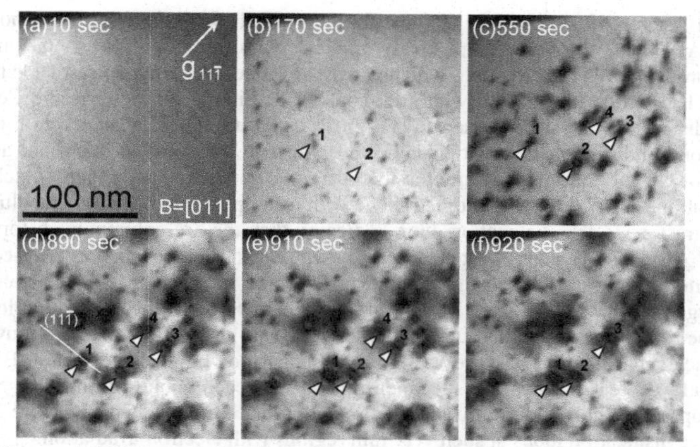

Figure 2. Sequential BF images for the nucleation-and-growth process of dislocation loops in CeO_2 under 200 keV electron irradiation at 300 K with an electron flux of 1.5×10^{23} electrons/m^2s. The micrographs (a)-(f) were taken at an identical region with changing electron irradiation time.

Atomic resolution STEM observations

Figure 3 shows a low magnification HAADF STEM image of CeO_2 taken from a [011] direction. Dislocation loops, formed under illumination of 200 keV electrons, are seen as dark line contrast with about 5 nm in size (indicated by arrows in Figure 3), which reveals that loops lie on {111} planes at edge-on conditions. It is seen that strong strain contrast is induced around the dislocation loops.

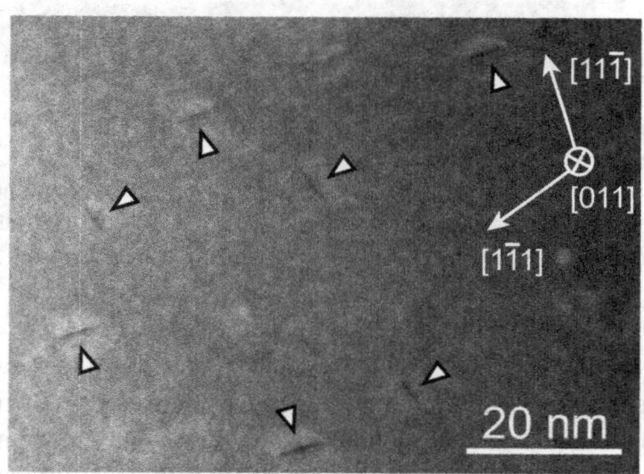

Figure 3. HAADF STEM image of dislocation loops in CeO_2 formed under 200 keV electron illumination during the observation at 300 K, showing dislocation loops on {111} planes at edge-on conditions (indicated by arrows).

Magnified images of an identical dislocation loop are shown in Figure 4(b) and (c), for HAADF and ABF STEM images, respectively, together with the atomic layer stacking in the fluorite structure from a [011] direction, which consists of nine-layers sequence of {111} planes (Figure 4(a)). The lattice image with white dot-contrast in Figure 4(b) reflects the Ce-column with a lattice distance of the (111) plane of 0.31 nm. A dislocation loop is located in Figure 4(b) with an edge-on condition at the position indicated with two arrows. It is seen that additional atomic layers of Ce-column is not inserted at the position of the dislocation loop, which reveals that the dislocation loop is not a stoichiometric one with a composition of CeO_2. The blunt image of lattice contrast near the dislocation loop indicates that the Ce-column is distorted. The corresponding ABF STEM image (Figure 4(c)) shows that the lattice image of Ce-column is severely distorted around the dislocation loop with dark and blunt contrast of Ce-column. An additional oxygen layer was, however, not observed at the position of the dislocation loop. This is probably due to the thick specimen thickness (about 50 nm) used in the present study, since a recent theoretical investigation has shown that the intensity of light elements in ABF STEM image drastically decreases when the specimen thickness exceed about 10 nm [11]. Since the dislocation loops are highly mobile during electron illumination as shown in Figure 2, it was unable to induce dislocation loops at such very thin regions with electron irradiation.

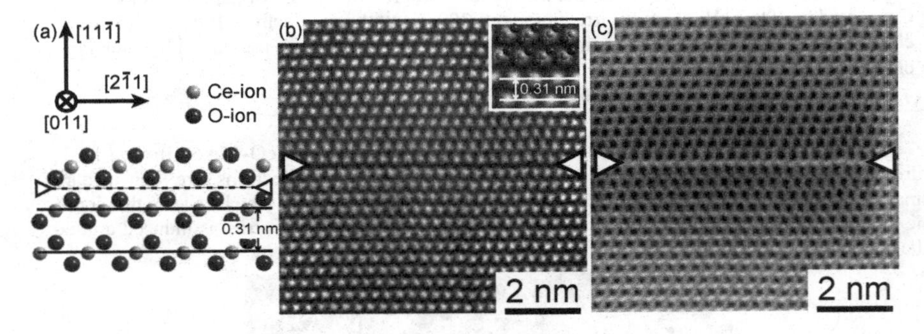

Figure 4. Atomic stacking of the fluorite structure from a [011] direction to show a nine -layers stacking sequence of {111} planes (a). High resolution HAADF STEM (b) and ABF STEM (c) images in CeO_2 taken from a [011] direction, including an identical dislocation loop (located between two arrows) formed under 200 keV electron irradiation. The corresponding atomic configuration is superimposed in a magnified image inserted in Figure 4(b).

Another example of HAADF STEM image containing a dislocation loop is shown in Figure 5. The dislocation loop lying on a $(11\bar{1})$ plane with an edge-on condition is seen to expand the $(11\bar{1})$ Ce lattice perpendicular to the habit plane of the loop. Further, strong distortion of Ce-lattice is induced around the dislocation loop compared to the one shown in Figure 4(b). The degree of distortion in Ce-lattice is found to depend on the individual dislocation loops. Such difference is presumably reflected to the difference in the existing depth of loops in the specimen.

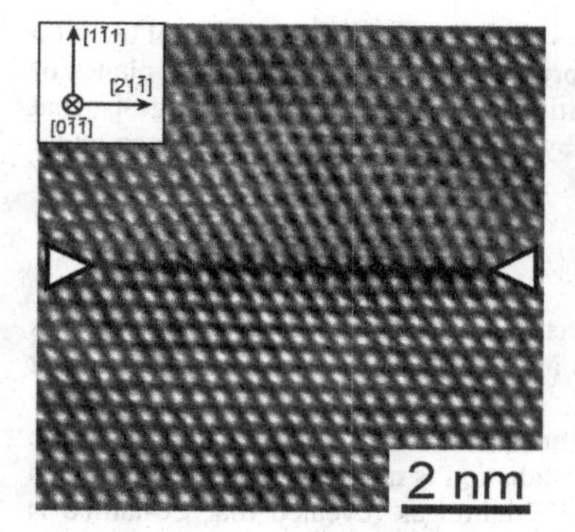

Figure 5. An example of high resolution HAADF STEM image of a dislocation loop in CeO_2 irradiated with 200 keV electrons. Strong distortion and expansion of Ce-column is seen around the dislocation loop compared to the one shown in Figure 4(b).

Lattice expansion of Ce-ion Columns

In order to obtain quantitative information of the lattice distortion, the signal intensity profile across dislocation loops was measured. Figures 6(a) and (c) are Fourier inverse transform images obtained from Fourier transformed images of Figure 4(b) and Figure 5, respectively, by using a mask containing transmitted spot and systematic (111) reflection spots. These processed images provide selective information relevant to {111} planes of Ce lattice. Figure 6(b) and (d) are the signal intensity profiles obtained from Figure 6(a) and (c), respectively, in which each dislocation loop is located between the plane 15 and 16 (denoted as P15 and P16). It is seen in Figure 6(a) and (c) that {111} planes of Ce-lattice is significantly distorted. The lattice spacing of {111} plane is strongly expanded to be 0.49 nm compared to the normal lattice spacing of 0.31 nm (Figure 6(d)), although the corresponding lattice spacing of the loop in Figure 4(b) reveals the same spacing of the normal value within an experimental error (Figure 6(c)). The background

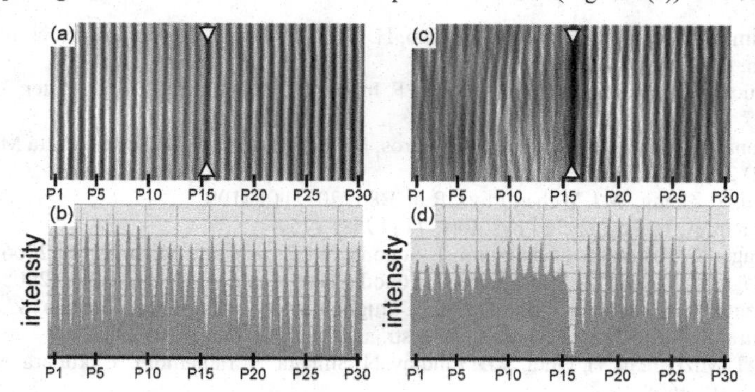

Figure 6. Fourier inverse transform images processed from HAADF STEM images containing a dislocation loop. Images (a) and (c) are processed from Figure 4(b) and 5, respectively, by using the transmitted spot and systematic {111} reflection spots. (b) and (d) are the intensity profiles of (a) and (c), respectively.

intensity increases near the dislocation loop and decreases significantly at the position of the loop (between P15 and P16). These results indicate the distortion and expansion of {111} planes of Ce-lattice are induced without an insertion of additional Ce-layers, which suggests that dislocation loop consist of oxygen interstitials induced by selective displacement damage under electron irradiation.

CONCLUSIONS

We have investigated the radiation-induced defects in CeO_2 irradiated with electrons through atomic resolution TEM techniques. Followings are drawn as conclusions in the present study.

High resolution STEM observations with HAADF and ABF imagings have shown that the dislocation loops in CeO_2 formed by 200 keV electron irradiation is not perfect dislocation loops with a stoichiometric composition. Analysis on the STEM images revealed that Ce-lattice is significantly distorted and expanded around dislocation loops. These results suggest that the non-stoichiometric dislocation loops formed under electron irradiation consist of oxygen interstitials.

ACKNOWLEDGMENTS

A part of electron microscopy observation/analyses was done at the HVEM Laboratory of Kyushu University. The authors are grateful to technical stuffs of both research facilities for their skillful technical assistance.

REFERENCES

[1] K.Yasuda, M. Nastasi, K.E. Sickafus, C.J. Maggiore, N. Yu, Nucl. Instr. and Meth. B136 (1998) 499.
[2] N. Sasajima, T. Matsui, K. Hojou, S. Furuno, H. Otsu, K. Izui, K. Murumura, Nucl. Instr. and Meth. B141 (1998) 487.
[3] M.A. Pouchon, M. Nakamura, Ch. Hellwig, F. Ingold, C. Degueldre, J. Nucl. Mater. 319 (2003) 37.
[4] G. Sattonnay, C. Grygiel, I. Monnet, C. Legros, M. Herbst-Ghysel, L. Thome, Acta Materia 60 (2012) 22.
[5] H.Y. Xiao, Y. Zhang, W.J. Weber, Phys. Rev. B86 (2012) 054109.
[6] W.A. McKinley, H. Feshbach, Phys. Rev. 74 (1948) 1759.
[7] K.Yasunaga, K. Yasuda, S. Matsumura, T. Sonoda, Nucl. Instr. and Meth. B250 (2006) 114.
[8] K.Yasunaga, K. Yasuda, S. Matsumura, T. Sonoda, Nucl. Instr. and Meth. B266 (2008) 2877.
[9] K. Shiiyama, T. Yamamoto, T. Takahashi, A. Guglielmetti, A. Chartier, K. Yasuda, S. Matsumura, K. Yasunaga, C. Meis, Nucl. Instr. and Meth. B268 (2010) 2980.
[10] H, Hojo, T. Mizoguchi, H. Ohta, S.D. Findlay, N. Shibata, T. Yamamoto, Y. Ikuhara, Nano Let. 10 (2010) 4668.
[11] S.D. Findlay, N. Shibata, H. Sawada, E. Okunishi, Y. Kondo, Y. Ikuhara, Ultramicroscopy 110 (2010) 903.
[12] A. Guglielmetti, A. Chartier, L.V. Brutzel, J.-P. Crocombette, K. Yasuda, C. Meis, S. Matsumura, Nucl. Instr. and. Meth. B 266 (2008) 5120.

Mater. Res. Soc. Symp. Proc. Vol. 1514 © 2013 Materials Research Society
DOI: 10.1557/opl.2013.356

Effect of Alloy Composition & Helium ion-irradiation on the Mechanical Properties of Tungsten, Tungsten-Tantalum & Tungsten-Rhenium for Fusion Power Applications

Christian E. Beck[1], Steve G. Roberts[1], Philip D. Edmondson[1] and David E. J. Armstrong[1]
[1]Department of Materials, University of Oxford, Parks Road, Oxford, OX1 3PH, UK

ABSTRACT

Model alloys have been made of pure W and 1% & 5% W-Ta and W-Re. Indentation hardness and modulus data were obtained by nanoindentation to assess the effect of composition on mechanical properties. Results showed that both the Ta and Re compositions hardened with increasing alloy content, greater in the W-5%Ta composition which showed an increase of 1.03GPa (17%), compared to a 0.43GPa (7%) increase in W-5%Re. The samples also showed very small increases in modulus of ~ 25GPa (6%) in both W-5%Re and W-5%Ta. The samples were implanted with 3000appm concentration of helium. All samples show a substantial increase in hardness of up to 107% in the case of pure W. An appreciable difference in modulus is also seen in all samples. Initial TEM work has shown no visible He bubbles, suggesting that the mechanical properties changes are due to He-vacancy cluster formation below the resolvable limit.

INTRODUCTION

Tungsten-based alloys are being considered as likely candidate materials for plasma-facing components in a nuclear fusion reactor. Structural components in the divertor and particularly the divertor plate will potentially be exposed to maximum operating temperatures in excess of 1000°C whilst being bombarded with 14.1MeV neutrons producing displacement damage of up to 120dpa over the divertor lifetime [1]. In addition to the cascade damage produced, helium retention in the first wall material from direct injection of plasma ions and neutron-induced transmutation events also has the potential to produce microstructural changes and in turn significant mechanical changes, although these are as yet not well characterised [2-4]. Tungsten-based alloys represent the greatest hope of being able to eventually fulfil the material requirements due to their high melting point (3695K), good resistance to sputtering and void swelling, relatively low tritium retention rates and good thermal resistance and conductivity properties, which remain stable under irradiation [5-6]. Tungsten alloys however have some drawbacks: high ductile to brittle transition temperature (DBTT), susceptibility to radiation embrittlement and processing issues due to their high melting temperature will have to be overcome [2,7-8].

The effect of alloying additions on the properties of tungsten is of interest for two reasons: firstly, deliberate alloying has the potential to improve material properties and secondly, to assess the effects of 'transmutation alloying'. Modelling has shown that over a 5 year service period for divertor conditions in a fusion power plant such as DEMO, over 5% of the tungsten would be transmuted, with rhenium and tantalum being two of the most significant transmutants [9]. It is therefore important that the effects of these alloying additions on the mechanical properties of model systems is understood both pre and post irradiation. There has also been speculation that tantalum like, rhenium and iridium, could have beneficial effects on the ductility of tungsten [10].

Due to the lack of a suitable facility to produce the level of neutron irradiation and helium implantation that a first-wall material would experience in a fusion power reactor and the inability to simulate reactor lifetime over laboratory time-scales, ion implantation techniques have been used to simulate the cascade and implantation damage [11-12]. Damage produced by ion implantation techniques is shallow – typically on the scale of a few hundred nm to several μm beneath the implanted surface. This necessitates the use of micromechanical testing techniques such as nanoindentation [12,13].

Helium damage is of particular interest as significant microstructural changes have been demonstrated in polycrystalline tungsten under helium implantation, potentially producing large changes in mechanical and thermal properties [2,6,14]. These changes can be substantially greater than those of self-ion implantation (cascade) damage alone [12]. Doses of up to 1200appm could be expected to be observed in first wall reactor materials [1]. Relatively little work has been done to date to characterise the effect of helium implantation on these alloys, either with or without concomitant displacement damage. This paper reports a nanoindentation study of the effects of helium implantation on the mechanical properties of W and dilute W-Re and W-Ta alloys.

EXPERIMENTAL DETAILS

Model alloys of 1 & 5% Re & Ta were produced by the arc-melting of high-purity elemental powders: 99.9% Ta & W, (Sigma Aldrich, USA) 99.99%Re (AEE, USA). The powders were weighed and mixed in a Turbula powder mixer (WAB, Switzerland) and compressed using a uniaxial press. The resultant pellets were melt-processed using a plasma-arc furnace in an argon atmosphere at the Department of Materials Science & Metallurgy at the University of Cambridge. The slugs produced were sliced into sheets and polished using diamond paste followed by a 50nm colloidal silica suspension to produce a high quality damage-free surface finish. Figure 1 shows typical microstructures as electron back-scatter diffraction (EBSD) inverse pole figure (IPF) maps, produced using Orientation Imaging Microscopy (OIM) software (EDAX-TSL, AMETEK, USA). In all materials, grains were equiaxed, of size ~50μm-500μm and without any significant texture.

Mechanical data were obtained from both irradiated and unirradiated samples by nanoindentation using an MTS NanoXP (Agilent, USA), using a Berkovich diamond indenter to a maximum depth of 2μm. The continuous stiffness measurement method (CSM) was used to evaluate the hardness and elastic modulus as a function of indentation depth without the need to run multiple load-unload cycles [15]. Grids of indents were placed over several grains for each alloy type.

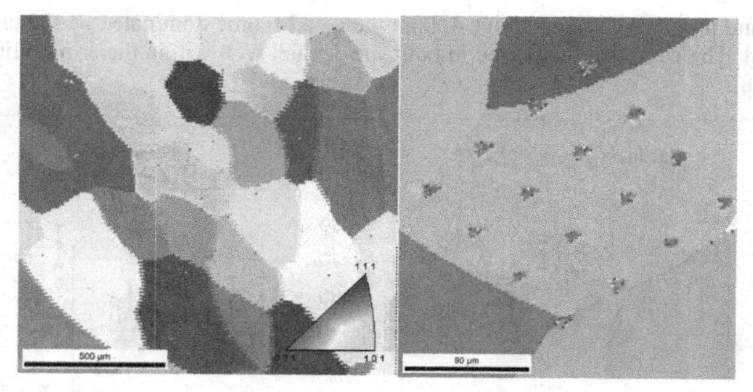

Figure 1 *- Inverse Pole Figure EBSD maps of surface normal direction of grain structure for a typical indentation site in pure W (unimplanted).*

The helium implantation was carried out at the National Ion Beam Centre, University of Surrey, using a 2MV tandem accelerator (HVEE, Netherlands) The implantation was carried out at ~300°C using a range of energies: 0.05, 0.1, 0.2, 0.3, 0.4, 0.6, 0.8, 1.0, 1.2, 1.4, 1.6 & 1.8 MeV, with fluences of 3.6×10^{15}, 1.5×10^{15}, 5×10^{15}, 1.0×10^{15}, 5.0×10^{15}, 5.0×10^{15}, 5.0×10^{15}, 5.0×10^{15}, 5.0×10^{15}, 5.0×10^{15}, 5.5×10^{15} and 6.0×10^{15} ions/cm^2 respectively, to produce a relatively flat implantation profile, as shown in figure 2. Profiles were calculated using SRIM (Stopping Range of Ions in Matter), assuming a displacement energy of 68eV [16]. Fluences were chosen to produce an implanted concentration of close to 3000appm over the implantation depth range, which in turn produced a roughly flat approximately 0.2dpa damage profile.

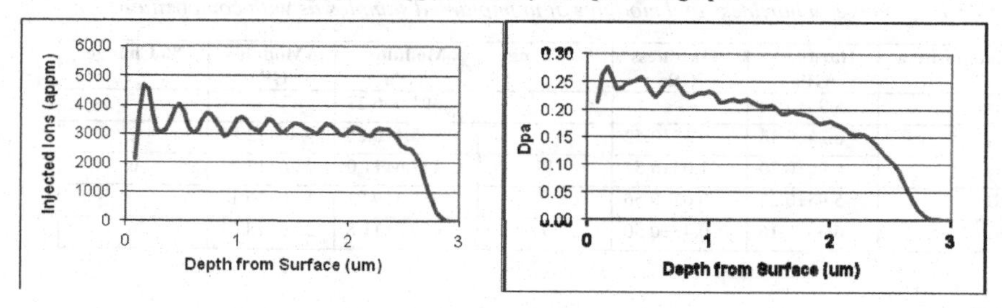

Figure 2 *– SRIM calculated He & displacement damage profiles in pure W for ~3000appm implantation*

RESULTS & DISCUSSION

Unimplanted alloys

Figure 3 shows indentation hardness and modulus values as a function of alloy composition. Values are shown as an average of a minimum of 16 indents over the 200-300nm indenter depth range. This range was chosen for later comparison with data of implanted materials from self-ion implantation experiments, in which, below 75nm indentation depth, the data are dominated by tip shape effects and initial pop-in and above 400nm the data become dominated by the unimplanted 'substrate' [12]. The compositions tested in both species fall well within their solubility limits in tungsten [17-18].

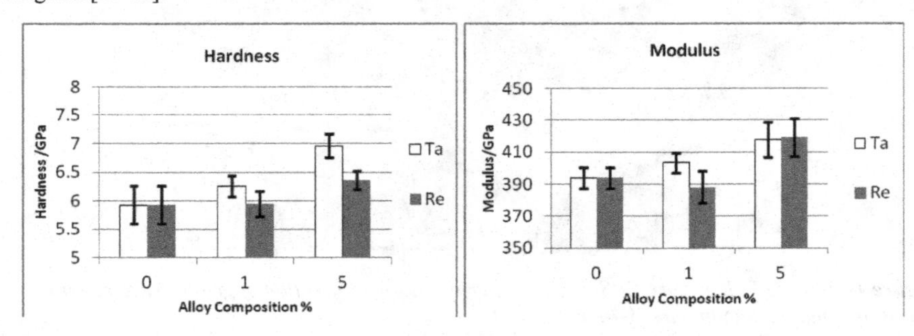

Figure 3 *- Indentation hardness & modulus for all W-Re & W-Ta compositions (error bars 1 standard deviation).*

The numerical hardness values alongside with the % change from pure W can be seen in table 1. Although there is slight trend of increasing hardness with composition in both W-Ta and W-Re alloys the only sample which shows a substantial increase is W-5%Ta, where the increase falls well outside of any potential error. The results also show a small increase in modulus at both 5% compositions, although this increase is negligible.

Table 1 *– Change in hardness and modulus of unimplanted samples as with composition.*

Composition	Hardness /GPa	ΔHardness /GPa	% Change	Modulus /GPa	ΔModulus /GPa	% Change
Pure	5.93±0.34	-	-	394.0±6.27	-	-
1Ta	6.26±0.18	0.33±0.52	6	403.4±6.28	9.40±12.55	2
5Ta	6.96±0.20	1.03±0.54	17	417.6±11.0	23.6±17.27	6
1Re	5.95±0.22	0.02±0.56	0	388.3±9.91	-5.70±16.18	-1
5Re	6.36±0.16	0.43±0.50	7	419.0±11.8	25.0±18.07	6

He Implanted

Figure 4 shows the modulus and hardness in the helium implanted materials compared with their unimplanted state, with the numerical values shown in table 2. It can be seen that the Ta alloys harden after implantation slightly more in absolute terms than the Re alloys. With the W-Re alloys the increase in hardness due to implantation increases with solute content , but this trend is reversed in the W-Ta alloys, with the W-5%Ta concentration sample hardening slightly less than the W-1%Ta alloy. The overall hardening effect is greater in the W-Ta alloys than the W-Re compositions. The proportional increase in hardening can be seen in table 2, with a reduction in the alloyed samples, especially in the W-5%Ta. The proportional change in the W-Re alloys remains approximately the same.

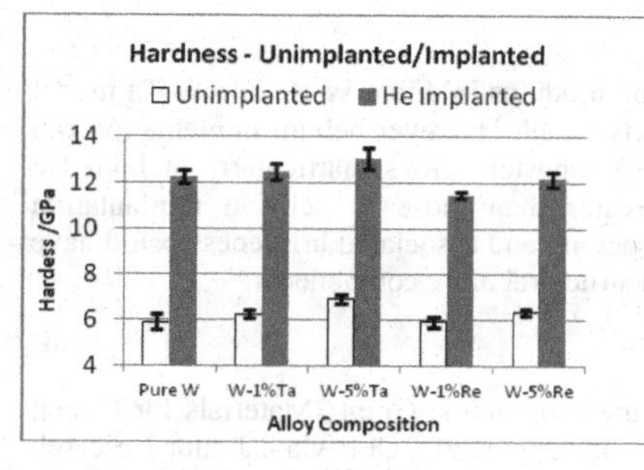

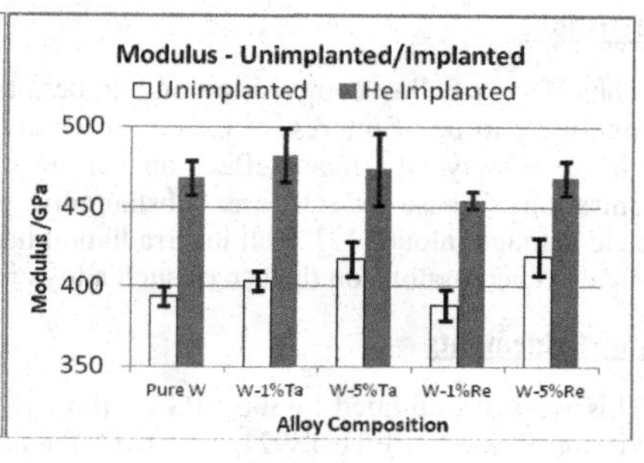

Figure 4 *- Indentation Hardness & Modulus for all W-Re & W-Ta compositions implanted with 3000appm He (error bars 1 standard deviation).*

Figure 4 and Table 2 also show that there is an appreciable increase in modulus of all the samples after implantation. As with the increase in hardness, the increases in modulus are marginally higher in the W-Ta alloys than the W-Re alloys, with a reduced increase in the W-5% alloy concentration samples in both the W-Re and W-Ta compositions. In all cases the increases are tangible and fall outside the range of experimental error

Table 2 – *Change in hardness and modulus from unimplanted to 3000appm He implanted*

Composition	ΔHardness /GPa	% Increase	ΔModulus /GPa	% Increase
Pure	6.36±0.64	107	74.4±16.67	19
1Ta	6.24±0.53	100	79.1±23.18	20
5Ta	6.13±0.65	88	56.1±33.10	13
1Re	5.56±0.39	93	66.5±15.41	17
5Re	5.85±0.46	92	49.4±22.20	12

Initial transmission electron microscopy (TEM) of the pure W sample implanted to 3000appm He revealed that no visible He bubbles had formed (figure 5). This indicates that the He in the sample exists as small He-vacancy clusters below the resolution limit of the TEM. Such small He-vacancy clusters may be the source of the observed increased hardening via He-vacancy cluster pinning of mobile dislocations. Further microstructural analysis is required to determine exactly how defect clusters may affect the mechanical properties of the samples

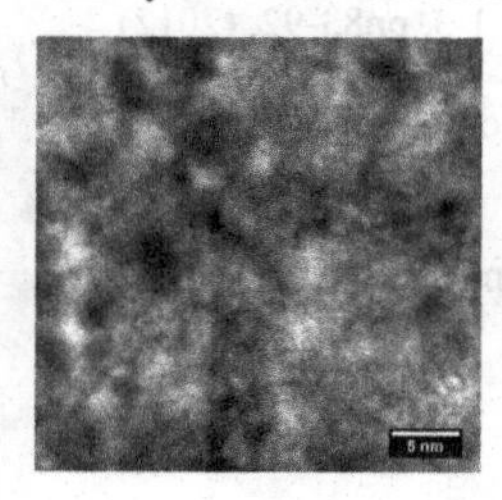

Figure 5 – *TEM micrograph of the 3000appm He implanted pure W. The image was recorded at ~1 μm underfocus. No Fresnel contrast was observed in a through-focus series indicating no He bubbles are present.*

Conclusions

The effects of alloy composition on hardness and modulus in dilute W-Re and W-Ta model systems likely to be of interest in fusion are relatively small. However helium implantation can have an extremely significant effect on hardness of tungsten alloys, particularly at high He concentrations. These effects are substantially greater than those of self-ion implantation (cascade damage) alone [12]. Helium-irradiation hardening and associated brittleness could have significant repercussions on the use of such alloys in structural alloy components

Acknowledgements

This work was funded by the EPSRC through the Programme Grant "Materials for Fusion and Fission Power" (EP/H018921/1). DEJA thanks the support of CCFE via a Junior Research Fellowship at St Edmund Hall, Oxford. CEB also thanks the Worshipful Company of Armours and Braziers for their support. We gratefully acknowledge the assistance of researchers and staff at the National Ion Beam Centre, University of Surrey, UK and the Department of Materials Science & Metallurgy at the University of Cambridge, UK.

References

1. H. Bolt, V. Barabash, W. Krauss, J. Linke, R. Neu, S. Suzuki & N. Yoshida, J. Nucl Mater, **329-333**, A, pp66-73, (2004)
2. N. Yoshida, J. Nucl Mater, **266-269**, pp197-206, (1999)
3. M. Baldwin & J. Doerner, Nucl Fusion, **48**, pp1-5, (2008)
4. Q. Xu, N. Yoshida and T. Yoshiie, Mater T JIM, **46**, 6, pp1255-1260, (2005)
5. P. Norajitra & 15 other authors, J. Nucl Mater, **367-370**, B, pp1416-1421, (2007)
6. R. Causey & T. Venhaus, Phys. Scr, **T94**, pp9-15, (2001)
7. N. Baluc, K. Abe, J.L. Boutard, V.M. Chernov, E. Diegele, S. Jitsukawa, A. Kimura, R.L. Klueh, A. Kohyama, R.J. Kurtz, R. Lässer, H. Matsui, A. Möslang, T. Muroga, G.R. Odette, M.Q. Tran, B. van der Schaaf, Y. Wu, J. Yu and S.J. Zinkle, Nucl Fusion, **47**, ppS696-S717, (2007)
8. M. Rieth & A. Hoffmann, Int. J. Refract. Met. H, **28**, pp679-686, (2010)
9. M. Gilbert & J-Ch. Sublet, Nuclear Fusion, **51**, 4, (2011)
10. M. Reith & 70 other authors, J. Nucl. Mater, **432**, 1-3, pp482-500, (2013)
11. M.B. Lewis, N.H. Packan, G.F. Wells & R.A. Buhl, Nucl. Instrum. Methods, **167**, pp233-247, (1979)
12. D.E.J. Armstrong, X. Yi, E.A. Marquisa & S.G. Roberts, J. Nucl Mater, **432**, 1-3, pp428-436, (2013)
13. C. D. Hardie & S. G. Roberts, J. Nucl. Mater, **433**, 1-3, pp174-179, (2013)
14. S. J. Zenobia, L. M. Garrison, G. L. Kulcinski, J. Nucl. Mater, **425**, 1-3, pp83-92, (2012)
15. W. C. Oliver & G. M. Pharr, J. Mater. Res, **7**, 1564-1583, (1992)
16. International A. ASTM E521 – 96, Standard Practice for Neutron Radiation Damage Simulation by Charged-Particle Irradiation, 2009, ASTM International, West Conshohocken, PA, (2009)
17. R. A. Ayres, G. W. Shannette & D. F. Stein, J. Appl. Phys, **46**, 4, pp1526-1530, (1975)
18. National Physical Laboratories, MTDATA Calculated Phase Diagram, http://resource.npl.co.uk/mtdata/phdiagrams/taw.htm

Synthesis, Characterization and Thermomechanical Properties

Mater. Res. Soc. Symp. Proc. Vol. 1514 © 2013 Materials Research Society
DOI: 10.1557/opl.2013.389

Atom-probe tomography of surface oxides and oxidized grain boundaries in alloys from nuclear reactors

Karen Kruska[1], David W Saxey[12], Takumi Terachi[3], Takuyo Yamada[3], Peter Chou[4], Olivier Calonne[5], Lionel Fournier[5], George D W Smith[1], Sergio Lozano-Perez[1]

[1] University of Oxford, Department of Materials, Oxford, United Kingdom.
[2] University of Western Australia, School of Physics, Perth, WA, Australia
[3] Institute of Nuclear Safety Systems Inc., Tsuruga, Fukui, Japan.
[4] EPRI, Palo Alto, CA, United States.
[5] Areva NP, Paris, France.

ABSTRACT

The preparation of site-specific atom-probe tomography (APT) samples containing localized features has become possible with the use of focused ion beams (FIBs). This technique was used to achieve the analysis of surface oxides and oxidized grain boundaries in this paper. Transmission electron microscopy (TEM), providing microstructural and chemical characterization of the same features, has also been used, revealing crucial additional information.

The study of grain boundary oxidation in stainless steels and nickel-based alloys is required in order to understand the mechanisms controlling stress corrosion cracking in nuclear reactors. Samples oxidized under simulated pressurized water reactor primary water conditions were used, and FIB lift-out TEM and APT specimens containing the same oxidized grain boundary were prepared and fully characterized. The results from both techniques were found fully consistent and complementary.

Chromium-rich spinel oxides grew at the surface and into the bulk material, along grain boundaries. Nickel was rejected from the oxides and accumulated ahead of the oxidation front. Lithium, which was present in small quantities in the aqueous environment during oxidation, was incorporated in the oxide. All phases were accurately quantified and the effect of different experimental parameters were analysed.

INTRODUCTION

Because of their excellent mechanical properties and corrosion resistance at operating temperatures, stainless steels (SSs) and Ni-based alloys are frequently used in pressurised water reactors (PWRs) [1]. With increasing years of operation, more incidences of stress corrosion cracking (SCC) were identified [2] (although most failures in stainless steels have been attributed to non-specification conditions). A major advance in the combat against SCC was the replacement of Alloy 600 with the more Cr-rich Alloy 690 in steam generator tubes [3].

Cracking in the commonly used stainless steels and Ni-based alloys is predominantly intergranular [4, 5], which suggests that preferential oxidation of grain boundaries (GBs) is an important part (not the only part) of the SCC mechanisms in these alloys. Although the crystal structure of these alloys is identical and they have similar mechanical properties, the ratios of the main alloying elements are different. This leads to different corrosion potentials at the surfaces in the same environment and potentially to the formation of different oxides at the surfaces and at GBs.

Previous studies on the surface oxides of 304-type SSs reported a double layer of an inner Cr-rich spinel oxide and an outer oxide consisting of magnetite-type spinel faceted nanocrystals [6, 7, 8]. There are also a number of studies of surface oxides on Alloy 690. Spinel oxides, mixtures of spinels and sesquioxides (M_2O_3, where M=metal) and even NiO were reported [9, 10, 11, 12, 13]. As some of the oxides found after autoclave testing in SS autoclaves were more Fe-rich than expected, it is possible that Fe originating from the autoclave steels went in solution and was subsequently incorporated in the oxides.

This study examines and compares the oxides formed on the surfaces and at GBs of 304 stainless steel (SUS304) and Alloy 690. To avoid Fe pick-up from autoclave steels, the Alloy 690 was tested in a Ti autoclave. Atom probe tomography (APT) and analytical transmission electron microscopy (ATEM) were used to study the oxide compositions and the microchemistry around the oxide regions with high spatial and chemical sensitivity. The implications of the findings for SCC mechanisms are discussed.

EXPERIMENT

Material

A 304 stainless steel (SUS304), solution treated at 1060 °C for 100 min and water-quenched, and a nuclear Ni-based alloy (Alloy 690) were tested (see Table I for composition).

Table I. Chemical composition of the studied alloys (at.%)

Alloy	Fe	C	Si	Mn	P	S	Ni	Cr	Mo	Cu	Al+Ti
SUS304	Bal.	0.19	0.62	1.6	0.005	0.002	8.8	19.6	< 0.01	-	-
Alloy 690	10.19	0.12	0.52	0.3	0.02	0.002	57.04	31.28	-	< 0.01	0.52

A coupon specimen was cut out from the SUS304 and one surface was polished to mirror finish with colloidal silica. Oxidation was performed at the INSS laboratories (Japan) in simulated PWR primary water (500ppm B, 2ppm Li and 30 cm^3 H/kg H_2O) in a stainless steel autoclave for 1500 hrs at 360°C. The Alloy 690 was polished to mirror finish with 1 μm diamond paste. Autoclave testing was performed at the AREVA laboratories (France) in simulated PWR primary water (1200 ppm B, 2 ppm Li and 43 cm^3 H/kg H_2O) in a Ti autoclave for 1500 hrs at 325°C. To simulate more aggressive conditions and to accelerate the oxidation process, the temperatures in both cases were chosen slightly above the usual 280-320 °C PWR primary side conditions.

Method

TEM and APT specimens were prepared with a focussed ion beam (FIB) lift-out method as described elsewhere [16]. APT samples were analysed in a LEAP® 3000X HR atom probe at 50 K using 532 nm wavelength 10 ps duration laser pulses to stimulate field evaporation, with a pulse energy of 0.6 nJ and a repetition rate of 200 kHz. (S)TEM analysis was performed using a JEOL 2100 operated at 200kV and JEOL 3000F operated at 300kV, both equipped with a Gatan Quantum GIF [14]. Energy-Filter TEM (EFTEM) acquisitions were performed with an incident half-angle of ~6mrad and a collection half-angle of ~22mrad. Multivariate statistical analysis

(MSA) was found to be a particularly effective method for reducing statistical noise and was used to process all EFTEM data shown in this paper [15].

RESULTS

Oxides in SUS304

Figure 1 shows a TEM bright field (BF) image and EFTEM maps acquired from a SUS304 lift-out specimen. The displayed region contains the oxidised surface (inner oxide) and an oxidised GB. The holes in the protective C layer and the surface oxide region at the top half of the sample are caused by FIB damage from the final thinning step. These regions were removed for the MSA and are therefore covered up in EFTEM maps (Figures 1b-e). The brighter vertical streak in the centre of the image is also a thickness effect from the FIB preparation. In the BF image Grain 1 appears darker, because it is near a zone axis, allowing the observation of its high density of dislocations.

The thickness of the surface oxide was measured to be 127±23 nm on Grain 1 and 98±10 nm on Grain 2. This is consistent with the measurements from 3D FIB slicing performed on this sample and reported in [17]. The GB oxide depth at this part of the GB is 680 nm, which is much deeper than the average oxide depth at this GB measured in the 3D FIB slicing data (234±109 nm). The oxide regions can be identified more easily in the EFTEM elemental map in Figure 1b.

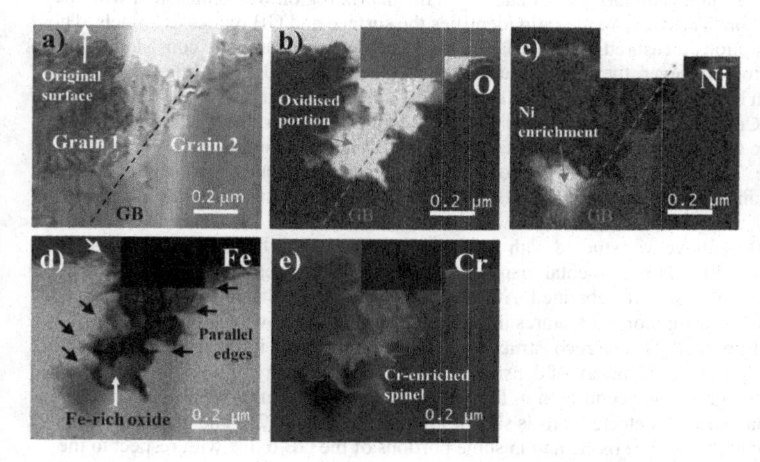

Figure 1. a) Cross-sectional TEM BF image of an oxidised GB in SUS304. b) EFTEM O map. c) EFTEM Ni map. d) EFTEM Fe map. e) EFTEM Cr map.

A thickness map was obtained with t/λ (thickness/mean free path) having an average value of ~1 in the matrix. The GB oxide in Figure 1b was 320±67 nm wide on average, it extended predominantly into Grain 1, but Grain 2 was also affected. The outer edges of the oxide portion had a zig-zag shape on both sides. The arrows in Figure 1d indicate some parallel edges of the oxide portion. Figure 1c shows Ni enrichment (relative to the bulk alloy) at the interface between the surface oxide and the matrix and to a limited extent at the edges of the GB oxide. A large Ni-rich particle was found at the oxide front. The Ni EFTEM map shows a gradual increase in the Ni concentration from the outer edges of this particle to its centre. The average Ni concentration in this particle was over 60 at.%. The projected dimensions of the Ni enriched particle were ~220×200 nm^2. Close consideration of its position in the Ni map (Figure 1c) reveals that the Ni enrichment is mostly confined to Grain 1.

EELS spectra were extracted from the EFTEM data set and the local compositions were analysed. The measured compositions are given in Table II.

Table II. The concentrations measured in the SUS304 EFTEM maps in Figure 1. Oxygen was omitted in the EFTEM quantification of the metallic features, as it can be attributed to surface oxidation of the TEM foil (6 at. % was measured).

Conc. (at.%)	**Fe**	**Cr**	**Ni**	**O**
Matrix	71±2	19±1	10±1	omitted
Surface Oxide	20±2	22±2	3±1	56±7
GB Oxide	16±3	27±2	3±1	54±7
Metallic Ni enriched region	36±2	4±1	61±10	omitted

The ratio of the main elements, Fe, Cr and Ni, in the matrix region were consistent with the nominal composition. The metal/oxide ratio identifies the surface and GB oxides as spinels. The O and Ni concentrations measured in the surface and GB oxides were mutually consistant within the margins of error. However, the Fe:Cr ratio in the surface oxide was 1:1.1, while the Cr content was much higher in the GB oxide (1:1.69). The Cr elemental map in Figure 1e shows that the absolute Cr concentration in the parts of the GB oxides is higher than in the surface oxide; parts of the GB oxide are bright, while there is no difference in contrast between matrix and surface oxide. The O map (Figure 1b) shows that the GB oxide extends beyond the Cr-rich regions (laterally into Grain 1). Part of the GB oxide seems to be more similar to the surface oxide.

The same GB as above was studied with APT. The samples were lifted-out as described in [16]. The marked region in the elemental map in Figure 2a shows an area representative of the region from which APT data was obtained. All ions of the data set are displayed in Figure 2b and a schematic with its most important features in Figure 2c. Figure 2d shows the individual ion maps generated from the APT data reconstruction. When looking at 3D ion maps in a 2D projection, it is important to be aware of density effects. A higher density of dots (ions) in the map can represent a greater concentration of the ion type or a higher density of all detected ions. An ion density map with all detected ions is shown in the left of Figure 2d. The map shows decreased density in the surface oxide and in some portions of the GB oxide with respect to the matrix. The density is increased, with respect to the matrix, in other portions of the GB oxide and in the dislocation oxide. Such a map showing all ions without chemical distinction is effective for showing crystallographic and microstructural features on the nanoscale, due to density differences and local magnification effects.

The APT data does not contain the centre of the GB oxide, but a portion from its left side. At the top of the data set is the surface oxide, below which lies the metallic matrix disrupted by oxidised dislocations running across the entire data set. At the bottom right of the APT data set the GB oxide comes into view. Although the analysed oxide is about 50 nm wide at the bottom of the data set, this is only a small part of the widespread GB oxide around this part of the GB, as the box in Figure 2a indicates. It is striking that the O/O_2, CrO/CrO_2 and FeO ions were all evaporated from core oxide regions, while the Fe_2O ions were only evaporated from the interface regions and the, probably more weakly oxidised, regions surrounding the oxidised dislocations. Fe_2O may have evaporated from regions in which oxides were only starting to grow, but no stoichiometric phase had yet formed at the end of the corrosion test. The last of the individual ion maps in Figure 2d shows the distribution of Li in the data set, which was present in all oxide regions. No B peak was discernible above the background noise of the mass spectrum for this sample.

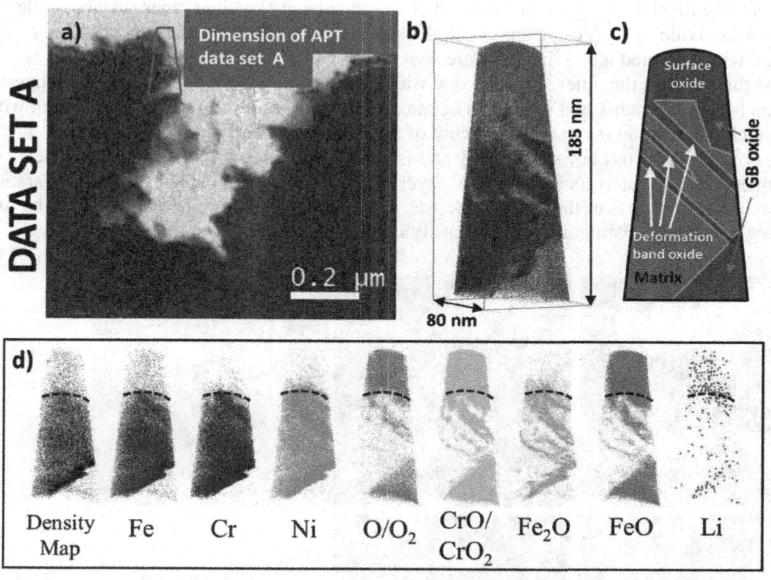

Figure 2. a) EFTEM oxygen map, showing a region representative of the SUS304 APT data set location. b) 3D APT atom map showing all ions of the data set, acquired from the SUS304 sample. c) Schematic showing the main features of the APT data set. d) Ion maps showing the distribution of the individual ion types within the data set. The dashed lines indicate the position of the interface between the surface oxide and the other features of the sample.

Selected regions of three APT needles were analysed to obtain information on the local compositions of the surface oxide, GB oxide, and the matrix regions. The average compositions of each region measured with APT can be found in Table III. Only two data sets contained GB oxides. The errors indicated in the table account for the variations between datasets. The concentrations of some lower concentration elements are also listed.

Table III. Compositions of the different features in the SUS304 sample, measured with the APT in at. %.

Conc.	Fe	Cr	Ni	Mn	Si	C	O
Matrix	68.2±0.7	19.6±1.2	9±0.4	1.4±0.2	0.49±0.01	0.07±0.03	0.7±0.1
Surf Ox	23.1±1.9	21.9±0.2	2.6±1.5	0.8±0.2	0.02±0.01	0.3±0.1	50.8±0.2
GB Ox	28.6±4.1	23.4±2.2	2.4±1.6	1.2±0.1	0.01±0.01	0.19±0.01	43.7±8.2

Oxides in Alloy 690

Figure 3 shows a TEM BF image and EFTEM maps acquired from an Alloy 690 lift-out specimen. The displayed region contains a double oxide layer (labelled inner surface oxide and outer surface oxide in Figures 3c and 3e) at the surface and a GB with minimal attack. Cr carbides, which formed along the GB were observed.

The thickness of the inner surface oxide was measured to be 16 ± 5 nm, with no noticeable variation between Grain 1 and Grain 2. The maximum oxide depth near the GB is 28 nm, which is not much deeper than the maximum depth of the inner oxide elsewhere in the sample. The thickness of the outer oxide varies greatly and is not easy to determine as it appears to be inhomogeneous and intermixed with the C layer, that was deposited during sample preparation. The maximum thickness of the outer oxide was measured to be 54 nm, above the GB. The two oxide regions can be distinguished more easily in the EFTEM Cr map in Figure 3e.

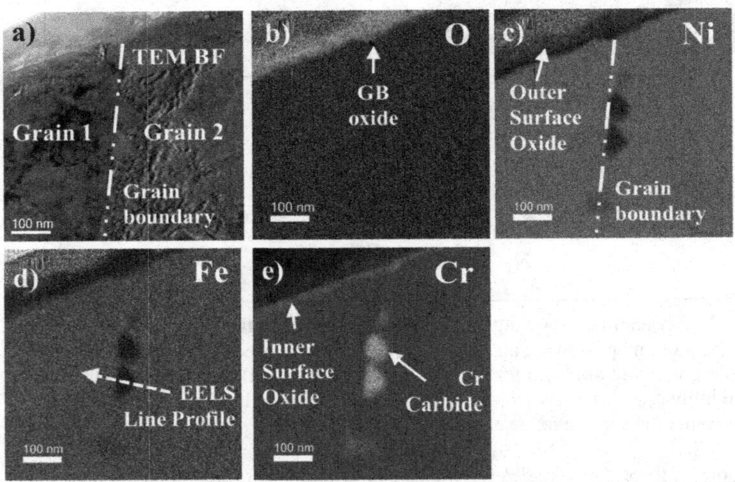

Figure 3. a) Cross-sectional TEM BF image of an oxidised GB in Alloy 690. b) EFTEM O map. c) EFTEM Ni map. d) EFTEM Fe map. e) EFTEM Cr map.

A thickness map was obtained with t/λ (thickness/mean free path) having an average value of ~1.1 in the matrix. The Cr carbide regions in Figure 3e were 58 nm and 52 nm wide respectively. They extend exclusively into Grain 2 and have faceted edges on the top.

EELS spectra were extracted from the EFTEM data set and the local compositions were analysed. The measured compositions are given in Table IV. EELS line profiling was performed across one of the Cr carbides (see dashed arrow in Figure 3d) and the Cr : C ratio was $63\pm3 : 37\pm3$.

Table IV. The concentrations measured in the EFTEM maps in Figure 3. Oxygen was omitted in the EFTEM quantification of the metallic features, as it can be attributed to surface oxidation of the TEM foil (4 at. % were measured).

Conc. (at.%)	Fe	Cr	Ni	O
Matrix	11 ± 1	32 ± 1	57 ± 1	omitted
Inner Surface Oxide	4 ± 1	30 ± 2	7 ± 4	58 ± 7
Outer Surface Oxide	3 ± 1	27 ± 2	16 ± 3	54 ± 7

The ratio of the main elements, Fe, Cr and Ni, in the matrix region were consistent with the nominal composition. The inner surface oxide was very thin and hard to separate from the matrix and outer oxide regions, both of which have a higher Ni concentration. This explains the large error on the Ni concentration and suggests that the real value is probably lower. The inner oxide seems to be depleted of Fe and Ni. The Cr:O ratio was measured to be ~1:2. The Fe concentration in the outer oxide was also low. The Ni:Cr:O ratio is close to 1:2:4.

The same GB as above was studied with APT. The samples were also lifted-out as described in [16].The marked region in the STEM high-angle annular dark field (HAADF) image in Figure 4a shows a region representative of the location where the APT needle was lifted out. All ions of the data set are displayed in Figure 4b, and a schematic of its most important features are displayed in Figure 4c. Figure 4d shows the individual ion maps generated from the APT data reconstruction. The density map shows a largely homogeneous density of ions across the dataset.

The APT data does not contain the centre of the GB oxide, but a portion from its right side. At the top of the data set is a small portion of the thin surface oxide. The left side of the data set is dominated by the slightly thicker oxide at the point where the GB intersects the surface. The right side of the data set contains the metal matrix.

The composition of the matrix and oxide were determined, and the results are shown in Table V. As only a single data set was analysed, no errors for variation between datasets could be given. The statistical (Poisson-) errors were negligible, due to the large number of atoms included in the evaluation. However, as explained above, there may be some systematic errors in the technique. If we assume an absolute error of 2 at.%, the measured composition of the matrix is consistent with the nominal composition within this margin.

The last of the individual ion maps in Figure 4d shows the Li distribution in the data set, which was present in all oxide regions. B was also found in this sample. However, the detected B was not limited to the oxide or metal regions and its origin could thus far not be determined.

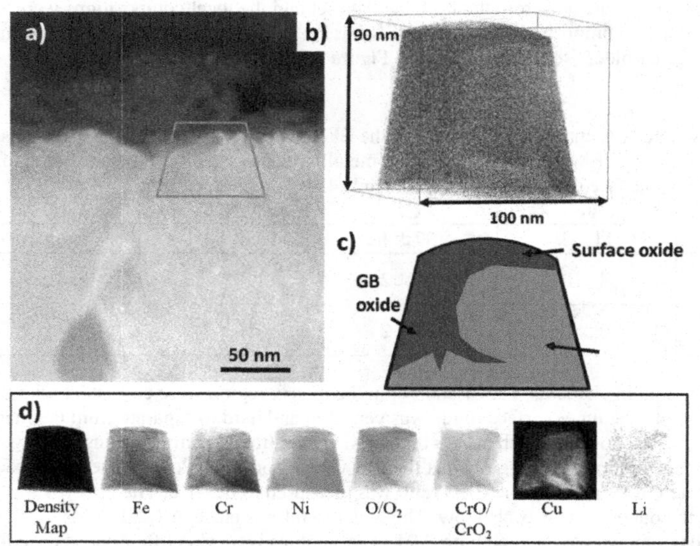

Figure 4. a) STEM HAADF image, showing a position representative of the region from which the APT data set was collected. b) All ions of the data set, acquired from the Alloy 690 sample. c) Schematic showing the main features of the APT data set. d) Ion maps showing the distribution of the individual ion types within the data set.

Table V. Compositions measured with APT. Co is listed with a question mark as the peaks at 29.5 m/e and 59 m/e overlap with NiH.

Element	Region	Ni	Cr	Fe	Al	Si	C	Mn	Cu	Co/NiH	Ti	O
Conc.(at. %)	Matrix	55.1	29.9	11.3	1.3	0.5	0.3	0.3	0.04	0.2	0.007	1.0
Conc.(at. %)	Inner Oxide	4.0	30.3	5.8	0.6	0.01	0.5	2.0	0.15	0.15	0.27	56.16

DISCUSSION

The surface and GB oxides were fully characterised for two different alloys. The combination of the different analysis techniques used for both specimens allows a detailed comparison between two alloys with similar properties but different compositions.

SUS304

As only slight preferential oxidation at GBs was expected in this alloy, the measured oxide depth of 680 nm was somewhat surprising. 3D FIB slicing performed in this sample type has shown that such a great oxide depth is untypical [17]. The facetted edges of the GB oxide suggest that the oxide has grown along defects intersecting the GB. The sample was not deformed and the surface was polished to a mirror finish with colloidal silica to minimise the damage in the surface region. This shows that even minor damages to the crystal structure in the surface region, can locally enhance oxide growth by up to an order of magnitude. The near-surface regions of operational components often contain many more defects than these laboratory samples.

Although the greater oxidation depth of the characterised GB is not representative for this material, the large oxide portion and its shape contain important information related to the oxidation mechanism:

1. The oxide is predominantly on one side of the GB.
2. The oxide front is blunt and the oxide is very wide-spread.
3. The oxide is not homogeneous and changes its composition along the GB.
4. The oxide has zig-zag shaped edges.
5. A Ni-rich region can be found in Grain 1 ahead of the oxidation front.

From these observations some conclusions about the oxide growth mechanism can be drawn:

1. Figure 1a shows many dislocations in Grain 1, whereas Grain 2 seems more uniform. Regions which contain many dislocations seem to oxidise preferentially.
2. Intuitively, one assumes that the GB is the fastest diffusion path for oxygen, however there is no sign of oxide growing into the GB beyond the dislocations at the sides. Rather it appears as if the oxidation front is terminated at a dislocation in Grain 1 that curls back towards the GB.
3. The Cr map in Figure 1e shows part of the oxide around the GB as being Cr rich. Part of the oxide looks porous, like a network of lines. It appears that the Cr-rich oxide has grown at first along the dislocations. The oxygen map in Figure 1b shows that the oxide regions go far beyond these Cr-rich (and Fe-depleted) parts. In the bottom centre of the GB oxide there is an area of Cr-depleted, Fe-rich oxide surrounded by Cr-rich lines, it seems that the Cr rich oxide at the oxidation front might form a barrier to further oxygen diffusion or at least slow it down, so that oxidation of the enclosed metal is favourable over diffusion deeper into the matrix.
4. The oxide has grown further in Grain 1 and forms pointy spikes. It appears that the dislocations, especially in Grain 1, act as much faster diffusion paths, than the GB itself. Therefore, the oxide is widespread, rather than limited to a narrow area around the GB.
5. Ni enrichment was observed ahead of the oxidation front. The entire oxide region is Ni depleted. It appears that Ni was rejected from the oxide, as it is nobler than Fe and Cr and can only be oxidised minimally at the present potential. It was also accommodated in the dislocation rich left grain, rather than in the GB. As Ni diffusion should happen by substitution, it appears that Grain 1 contains a higher number of vacancies, probably in dislocations, than the GB.

The compositional analysis of the matrix and surface oxide performed with APT and EFTEM was mostly in good agreement (see Tables II and III), except for the consistently low values for oxygen in the APT data. This is commonly observed in this type of oxide and is discussed in [16]. Variations in the Fe, Cr and also O concentrations in the GB oxide, that become apparent in larger errors, particularly in the APT data, account for the inhomogeneity of the oxide. Comparison of the EFTEM O (Figure 1b) and Cr (Figure 1e) maps illustrates a Cr-rich (former dislocation) substructure within the oxide. It is likely that the oxide portion in the APT data contains more of the Fe-rich oxide between the Cr-rich regions around the dislocations.

Alloy 690

No or minimal preferential attack at GBs was observed in Alloy 690 in TEM studies [18], making it an excellent replacement for Alloy 600 in PWR environments. This behaviour was confirmed in the present study.

In the literature, oxide films on Alloy 690 were characterised with XPS and AES depth profiling [9, 12], SIMS and ESCA [18]. Cr_2O_3 and $NiCr_2O_4$ or a combination of both were most commonly reported in the inner oxide layer [18, 19], but some studies also report NiO and hydroxides in the outer oxide layer [18]. A very detailed and thorough TEM and SEM study has clearly shown a triple layer structure of Cr_2O_3 particles, a continuous $Ni_{(1-x)}Fe_xCr_2O_4$ intermediate layer and an outer layer of facetted $Ni_{(1-z)}Fe_{(2+z)}O_4$ crystals and hydroxide particles [10].The last report shows HR TEM images and the respective Fourier transform diffractograms and identifies all phases without ambiguity.

However, the results presented in this paper suggest a double layer structure of a Cr_2O_3 inner oxide and an outer $NiCr_2O_4$ spinel. The reason for the different results despite similar testing conditions may lie in the autoclave material. The study by Sennour et al. [10] probably uses a stainless steel autoclave. The ion-exchange resin included in the recirculation loop used in the study was not operated under optimum conditions, allowing a higher level of metallic cations in the water. Fe from the stainless steel walls of the autoclave could go into solution and was subsequently incorporated in the outer oxide layers on the Alloy 690. This could not happen in this study, as a Ti autoclave was used. However, small amounts of Ti were incorporated in the inner oxide (see Table V) and quite possibly in the outer oxide, the latter will have to be confirmed in further experiments.

Comparision: SUS304 vs Alloy 690

It is apparent that the oxide formation in these two alloys is very different. While the original sample surface is preserved for SUS304 and oxide formation occurs by diffusion (inner layer) and deposition (outer layer), there was no evidence of the position of the original surface for Alloy 690. It appears that the original surface in the case of the Alloy 690 erodes until a thin oxide film has formed, preventing further oxidation. While diffusion is enhanced at GBs and leads to enhanced attack in SUS304, this does not seem to be the case for the oxide growth in Alloy 690. The passivating layer on SUS304 is a chromite-type spinel, while the passivating layer on Alloy 690, where less Fe is available, is a Cr_2O_3 sesquioxide. Both oxides contained small amounts of Li from the PWR water. It is interesting, that the GB attack "by diffusion" in SUS304 came to a halt, when a Ni particle with a similar Ni concentration as in the Alloy 690 had formed. It is possible, that the prevalent oxidation mechanism changes at this point and further diffusion is stopped. A contributing factor may be that the vacancies in this area were used up by Ni diffusion, which slows down O diffusion and hinders further oxidation.

CONCLUSIONS

The surface and GB oxides of two alloys were fully characterised and their structure and chemistry was compared. A number of conclusions could be drawn from these experiments:

1. Li (and potentially other trace elements) can be detected with APT. Li is always incorporated in the oxide phase, in both, the steel and the Ni-based alloy.
2. The oxide structure on SUS304 oxidised in simulated PWR primary conditions consists of a double layer. The inner passivating oxide layer is a $FeCr_2O_4$-type spinel. The outer layer was not analysed in this study.

3. The oxide structure on Alloy 690 formed in simulated PWR primary water conditions (in absence of steel) also consists of a double layer. The inner passivating oxide layer is a Cr_2O_3-type sesquioxide. The outer layer consists of inhomogeneous islands of $NiCr_2O_4$-type spinel.
4. There is almost no GB attack in the Alloy 690 and the GB attack in the SUS304 comes to a halt at a Ni enriched "particle" which has formed by the oxidation process itself. A similar mechanism may be responsible for this effect.
5. Oxide compositions can be reliably measured with a combination of EELS and APT. While the quantification of O is difficult with APT, the ratio between the alloying elements can be measured accurately. The comparison with EELS mapping sets the data into a more complete picture and helps to reliably identify oxide phases.

The methodology used in this paper can be used to analyse the growth of surface oxides on other metals and alloys and help in understanding oxide structures and growth mechanisms.

ACKNOWLEDGEMENTS

The authors thank INSS, EPRI and AREVA for the provision of samples and funding.

REFERENCES

[1] P M Scott. *Corrosion Science*, 25:583 – 606, 1985.

[2] P M Scott. *Corrosion issues in light water reactors : stress corrosion cracking*, An overview of materials degradation by stress corrosion in PWRs, pages 3 – 24. IOM3, 2007.

[3] P Diano, A Muggeo, J C Van Duysen, and M. Guttmann. *Journal of Nuclear Materials*, 168:290 – 294, 1989.

[4] S P Lynch. *Acta Metallurgica*, 36:2639 – 2661, 1988.

[5] V Y Gertsman and S M Bruemmer. *Acta Materialia*, 49:1589 – 1898, 2001.

[6] T Terachi, K Fujii, and K Arioka. *Journal of Nuclear Science and Technology*, 42:225 – 232, 2005.

[7] S M Bruemmer, E P Simonen, P M Scott, P L Andresen, G S Was, and J L Nelson. *Journal of Nuclear Materials*, 274:299 – 314, 1999.

[8] S M Bruemmer and L E Thomas. *Surface and Interface Analysis*, 31:571 – 581, 2001.

[9] I Betova, M Bojinov, V Karastoyanov, P Kinnunen, and T Saario. *Corrosion Science*, 58:20 – 32, 2012.

[10] M Sennour, L Marchetti, F Martin, S Perrin, R Molins, and M Pijolat. *Journal of Nuclear Materials*, 402:147 – 156, 2010.

[11] J Huang, X Wu, and E-H Han. *Corrosion Science*, 51(12):2976 – 2982, 2009.

[12] F Huang, J Q Wang, E H Han, and W Ke. *Journal of Materials Science & Technology*, 28:562 – 568, 2012.

[13] X Li, J Wang, E-H Han, and W Ke. *Corrosion Science*, In press 2012.

[14] S Lozano-Perez, V de Castro Bernal, and R J Nicholls. *Ultramicroscopy*, 109:1217–1228, 2009.

[15] S Lozano-Perez. *Journal of Physics: Conference Series*, 126, 2008.

[16] K Kruska, S Lozano-Perez, D W Saxey, T Terachi, T Yamada, and G D W Smith. *Corrosion Science*, 63:225 – 233, 2012.

[17] S Lozano-Perez, K Kruska, I Iyengar, T Terachi, and T Yamada. *Corrosion Science*, 56:78–85, 2012.

[18] F Carrette, M C Lafont, G Chatainier, L Guinard, and B Pieraggi. *Surface and Interface Analysis*, 34:135–138, 2002.

[19] P Combrade, P M Scott, M Foucault, E Andrieu, and P Marcus. In *Proceedings of the Twelfth International Conference on Environmental Degradation of Materials in Nuclear Power Systems-Water Reactors*, pages 883–890, 2005.

Mater. Res. Soc. Symp. Proc. Vol. 1514 © 2013 Materials Research Society
DOI: 10.1557/opl.2013.140

Micromechanical testing of oxidised grain boundaries in Ni Alloy 600

Alisa Stratulat[1] and Steve G. Roberts[1]
[1]Department of Materials, University of Oxford, UK

ABSTRACT

Micromechanical testing of focused ion beam (FIB) machined cantilevers was used to study oxidised grain boundaries in Ni-alloy 600. The Ni-alloy 600 samples were exposed in simulated PWR primary water at 325°C for 4500h with a hydrogen partial pressure of 30kPa. The FIB was used to machine small cantilever beams at the selected sites in the Ni alloy 600, cut so that the beam contained a selected grain boundary close to the built-in end. The FIB was also used to make a pre-crack, 700 nm deep, on the grain boundary. Cantilevers were loaded at the free end using a nanoindenter. Cantilevers milled in the un-oxidised sample yielded, and did not fracture. The specimens containing oxidised grain boundaries fractured at the boundary after small amounts of plasticity. Load vs. displacement data were used to calculate the fracture toughness of the oxidised grain boundaries. The fracture toughness associated with fracture of grain boundary oxide for these cantilevers was in the range 0.73-1.82MPa $(m)^{1/2}$, with an average value of 1.3MPa $(m)^{1/2}$. We believe this to be the first time the fracture toughness of an oxidised grain boundary has been determined.

INTRODUCTION

Stress Corrosion Cracking (SCC) involves a highly complex interplay of diffusional, chemical and mechanical factors in a series of related mechanisms, and affects material performance in a wide range of materials systems and environments. Within the nuclear industry where safety and structural integrity are crucial concerns SCC is of great significance; in particular in some Ni based alloys, and in austenitic stainless steels. In such systems SCC is predominantly along grain boundary paths. Alloy 600 is a solution strengthening alloy of composition 75% Ni, 15% Cr and 7%Fe [1]. Ni Alloy 600 mainly fails intergranularly in the pressurized water reactor (PWR) environment [2] in the steam generator tubes [3-4]. One of the causes of intergranular cracking is the degradation of grain boundaries by oxidation. This study will describe a method recently developed to determine the fracture toughness of individual oxidised grain boundaries in Ni Alloy 600, following the work of Di Maio and Roberts on microcantilever fracture of silicon and WC coatings [5], and that of Armstrong, Wilkinson and Roberts on fracture of bismuth embrittled grain boundaries in copper [6].

EXPERIMENTAL DETAILS

Ni Alloy 600 samples provided by AREVA were first polished down to a 1μm diamond finish and then exposed in PWR primary water at 325° C for 4500h with a hydrogen partial pressure of 30kPa. Lastly, the samples were polished with colloidal silica, which also revealed the intergranular oxide. Oxidised grain boundaries were examined by cutting FIB trenches and the ones perpendicular to the surface were selected for cantilever production. FIB was used to machine small, pentagonal cross-section cantilevers (5μm wide by 25μm long), cut so that the beam contained, close to the build-in end, a selected grain boundary as in Figure 1. Cantilevers were pre-notched at the grain boundary using a FIB single cut of 700nm depth.

An MTS Nanoindenter XP System was used to perform AFM-type scans to image the cantilevers, and then to load them close to the free end. Load vs. displacement data were used to calculate the fracture toughness of different oxidised grain boundaries.

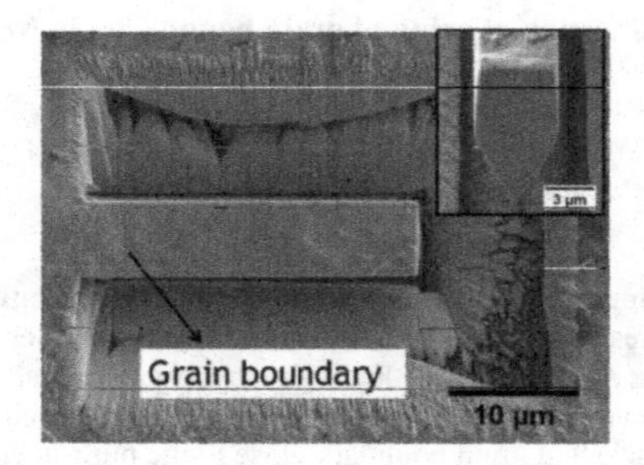

Figure 1. FIB machined cantilever in Ni alloy 600 with a pentagonal cross section.

Fracture toughness was calculated (as by Di Maio[5]) using the dimensions of the cantilever and the load at which the cantilever failed, via the following equation:

$$K_{I_c} = \frac{PLy}{I}\sqrt{\pi a}F\left(\frac{a}{b}\right) , \tag{1}$$

where P is the load applied by the nanoindenter, L is the length from the grain boundary to the loading point, a is the depth of the notch and I is the moment of inertia of the beam cross section. I is calculated as:

$$I = \frac{wb^3}{12} + \left(y - \frac{b}{2}\right)^2 bw + \frac{w^4}{288} + \left[\frac{b}{6} - (b - y)\right]^2 \frac{w^2}{4}, \tag{2}$$

where w and b are the dimensions shown in Figure 2 and y is defined as:

$$y = \frac{\frac{b^2 w}{2} + \frac{w^2}{4}\left(b + \frac{w}{6}\right)}{bw + \frac{w^2}{4}} \tag{3}$$

F given in Eq. (1) is approximated to be:

$$F\left(\frac{a}{b}\right) = 1.85 - 3.38\left(\frac{a}{b}\right) + 13.24\left(\frac{a}{b}\right)^2 - 23.26\left(\frac{a}{b}\right)^3 + 16.8\left(\frac{a}{b}\right)^4 \tag{4}$$

The cantilevers were imaged using a Scanning Electron Microscope and the fracture surface was observed.

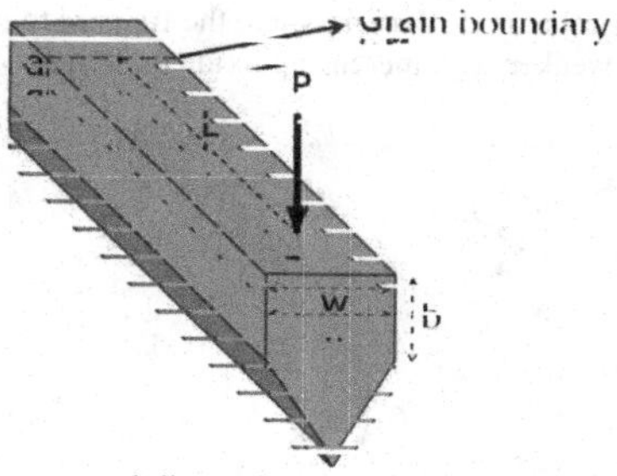

Figure 2. Geometry and dimensions of the cantilevers used in the study.

RESULTS

Sample exposed for 4500h

Load vs. displacement curves for two different cantilevers are presented in Figure 3. The load vs. displacement curve indicates fracture of the cantilevers, which is confirmed by micrographs of the cantilevers that were taken after the tests, shown in Figure 4.

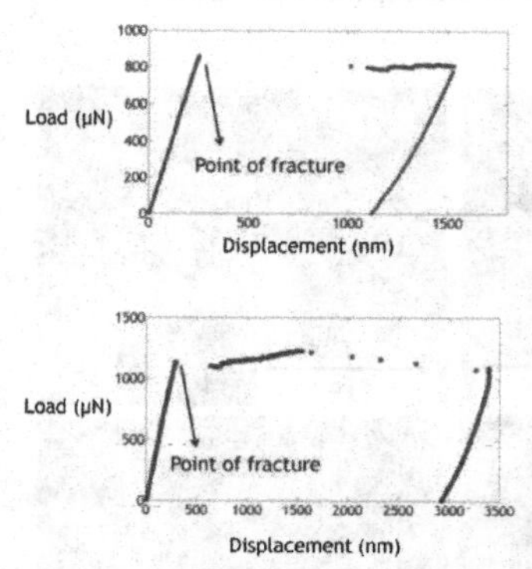

Figure 3. Load vs. displacement data for oxidised grain boundaries showing fracture, a rapid displacement increase at a critical load.

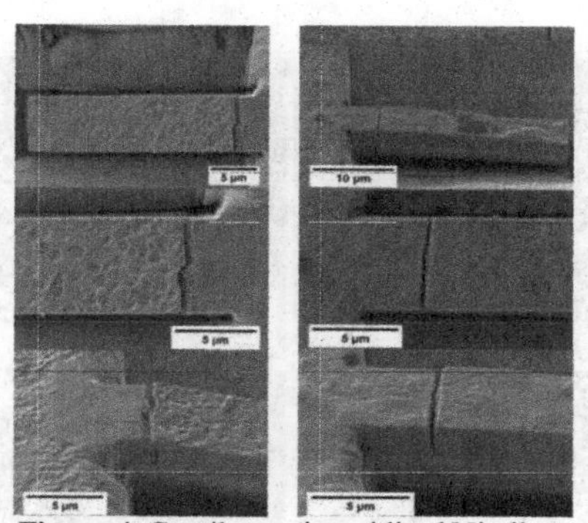

Figure 4. Cantilevers in oxidised Ni-alloy, after testing, showing fracture at the oxidised grain boundary.

The results for 22 tested cantilevers are presented in Figure 5. The fracture toughness associated with fracture of grain boundary oxide for these cantilevers was in the range of 0.73 – 1.82MPa $(m)^{1/2}$, with an average value of 1.3MPa $(m)^{1/2}$. The thickness of the intergranular oxide was measured using the Scanning Electron Microscope. It can be observed from Figure 5 that there is a weak correlation between the oxide thickness and the fracture toughness value, such that the grain boundaries become weaker with increasing oxide thickness.

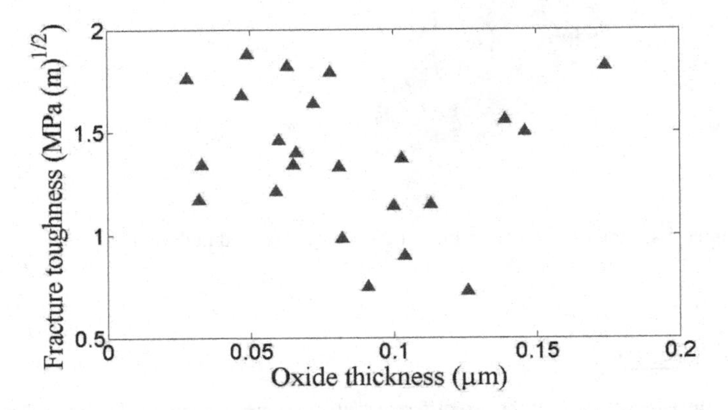

Figure 5. Calculated fracture toughness for tested cantilevers vs. oxide thickness.

Fracture surfaces were observed using the Scanning Electron Microscope and are shown in Figure 6.

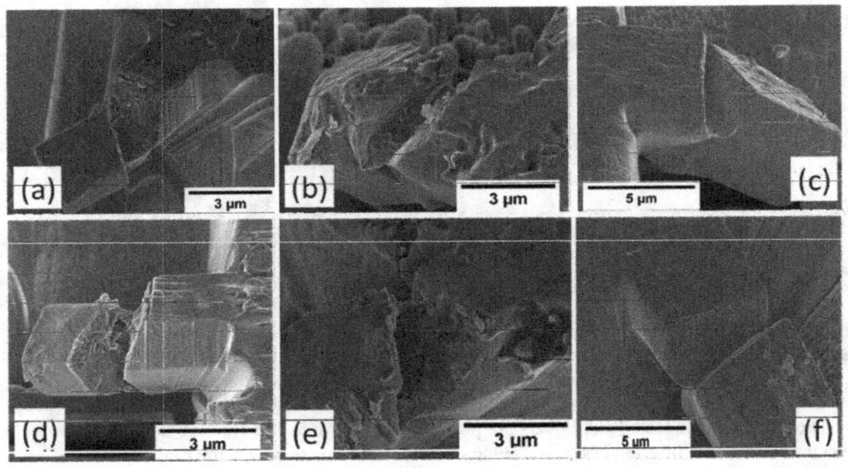

Figure 6. A discussion of different fracture surfaces at different toughness values. The fracture toughness values, are: (a, d) 0.73MPa $(m)^{1/2}$, (b, e) 1.64MPa $(m)^{1/2}$ and (c, f) 1.82MPa $(m)^{1/2}$.

Un-oxidised sample

Load vs. displacement curve for one of the six tested cantilevers made in un-oxidised Ni-alloy 600 is presented in Figure 7. The load vs. displacement curve is typical for a plastically deformed cantilever. The curve has initial elastic loading, yield and then plastic deformation. Figure 8 shows a micrograph of such a cantilever after testing: no fracture has occurred, but slip bands are present.

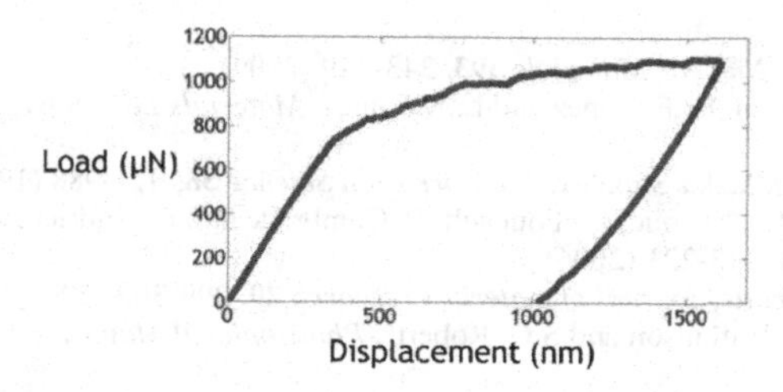

Figure 7. Load vs. displacement data for an un-oxidised cantilever, showing no fracture.

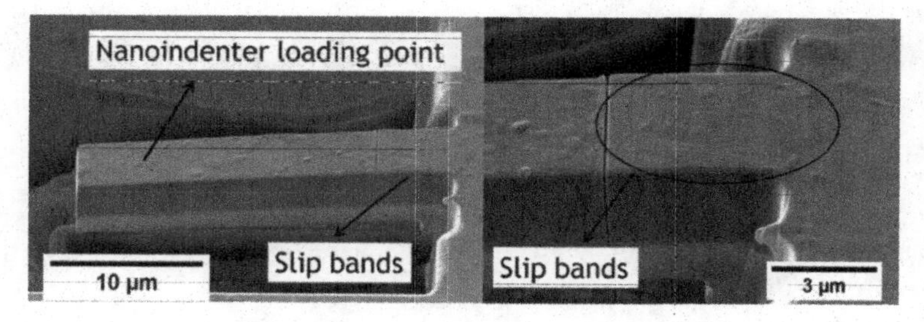

Figure 8. Cantilever after testing in un-oxidised Ni-alloy, showing slip bands (indicated by the arrow).

CONCLUSIONS

This study is believed to be the first that determines the fracture toughness of an oxidised grain boundary. The method described is a valid testing method for determining grain boundary strength in SCC susceptible alloy. Cantilevers milled in un-oxidised specimens yielded, and did not fracture. The specimens containing oxidised grain boundaries fractured at the boundary after small amounts of plasticity. The fracture toughness associated with fracture of grain boundary oxide for tested cantilevers was in the range of 0.73 – 1.82MPa $(m)^{1/2}$, with an average value of 1.3MPa $(m)^{1/2}$.

ACKNOWLEDGEMENTS

The author would like to thank to D.E.J. Armstrong for his help in exposing the fracture surfaces, Olivier Calonne and Marc Foucault from AREVA for providing the specimens and for useful suggestions and discussions. The research was supported by AREVA and the Engineering and Physical Science Council (EPSRC).

REFERENCES

1. R. S. Dutta, *Journal of Nuclear Materials* **393**, 343-349 (2009).
2. A. Aguilar, J. L. Albarran, H. F. Lopez and L. Martinez, *Materials Letters* **61**, 274-277 (2007)
3. R. B. Rebak and Z. Szklarska-Smialowska, *Corrosion Science* **38**, 971-988 (1996)
4. J. Panter, B. Viguier, J. M. Cloue, M. Foucault, P. Combrade and E. Andrieu, *Journal of Nuclear Materials* **348**, 213-221 (2006)
5. D Di Maio and SG Roberts, *Journal of materials research* **20**, 299-302 (2005)
6. D.E.J. Armstrong, A.J. Wilkinson and S.G. Roberts, *Philosophical Magazine Letters* **91**, 394-400 (2011)

Mater. Res. Soc. Symp. Proc. Vol. 1514 © 2013 Materials Research Society
DOI: 10.1557/opl.2013.448

Elastic strains in polycrystalline UO_2 samples implanted with He: micro Laue diffraction measurements and elastic modeling

Axel RICHARD[1], Etienne CASTELIER[1], Herve PALANCHER[1], Jean-Sebastien MICHA[2], Philippe GOUDEAU[3]

[1] *CEA, DEN, DEC/SESC, Centre de Cadarache, 13 108 St Paul lez Durance, France*
[2] *CEA, INAC, 38 054 Grenoble Cedex 9, France*
[3] *Institut Pprime, CNRS-Universite de Poitiers-ENSMA, BP 30179 - 86 962 Futuroscope, France*

Abstract :
In the framework of the study of long-term storage of the spent nuclear fuel, polycrystalline UO_2 samples have been implanted with He ions. The thin implanted layer, close to the free surface is subjected to elastic stresses which are studied by x-ray diffraction (micro Laue diffraction) and a mechanical modeling. A simple expression of the displacement gradient tensor has been evidenced; it concerns only three terms (ε_3, ε_4 and ε_5) which strongly evolve with considered grain orientations. Finally, we show that results obtained with micro diffraction are in very good agreement with conventional x-ray diffraction measurements done in laboratory at macro scale.

Keywords : Ion implantation, x-ray micro diffraction, strains measurements, modeling

INTRODUCTION

Majority of water reactors use uranium dioxide (UO_2) as their fuel. Fission of ^{235}U leads produces highly energetic fission products that move rapidly and deposit their kinetic energy throughout the UO_2 fuel pellet. The high kinetic energy of the fission products is redistributed throughout an extended spatial region, resulting in a significant increase in the temperature of the fuel pellet. It is this heat deposited in the pellet that is used to heat the reactor coolant, water, and ultimately used to generate steam that drives the turbines to produce the electricity at a nuclear power station [1]. Complex physical phenomena take place in UO_2 nuclear fuel in nuclear reactor and then in storage condition. One of the main important phenomena in the studies of the nuclear fuel spent is the structural swelling which is partially initiated by the alpha decay (helium production) of radioactive fission products created during irradiation. For studying this effect, as-prepared UO_2 samples are implanted with He ions which leads to the formation of a thin layer at the top surface of the sample. This approach allows easy handling of low radioactive samples while simulating storage conditions.

X-ray diffraction is used to measure induced strains in the implanted layer: monochromatic macro-diffraction and polychromatic micro-diffraction in Laue mode. Measured strain tensor is validated thanks to a mechanical model.

EXPERIMENTAL DETAILS

Sample preparation

In this study, four polycrystalline UO_2 disks (8.2 mm in diameter, 1 mm thick) have been prepared following a standard process [2] which limits surface defects and maintains an O/U stoichiometric ratio close to 2. Average grain diameter is about 18 μm. These samples have been implanted with 60 keV He ions at different fluencies ranging between 10^{15} and $2x10^{16}$ ions/cm^2.

The implantation depth is estimated to be 0.23±0.1 µm (SRIM code calculation). For the sake of clarity, the work described here is restricted to the study of the sample implanted at a fluency of 10^{16} ions/cm^2. A schematic view of the implantation is shown figure 1.

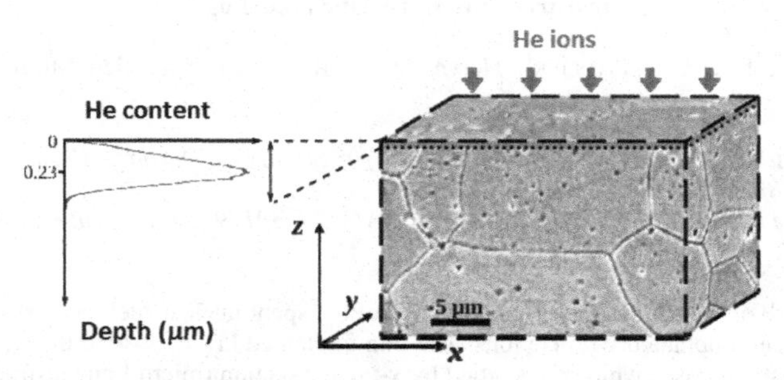

Figure 1 : He ions are implanted in a thin layer at the top surface of the sample.

Macro diffraction measurements

The first analysis has been done using a Bruker D8 diffractometer equipped with a copper x-ray source ($CuK_{\alpha 1}$ and $CuK_{\alpha 2}$ wavelengths). The use of an x-ray beam with macroscopic size combined with sample rotation (along a perpendicular axis to the sample surface) allows probing a large number of grains (about 10^6) at the same time. Diffraction diagrams have been recorded in θ/2θ mode for as-prepared and implanted samples in [10–120] 2θ angular range. Figure 2 shows an enlargement of these diagrams around the [311] Bragg peak.

Since the x-ray penetration depth (around 2µm) is greater than the implanted layer thickness, this technique probes simultaneously the strain free substrate and the implanted layer. Consequently, the [311] diffraction peak is split as shown on the diffraction pattern figure 2: the more intense diffraction peak component results from the free strain substrate (as-prepared state) whereas the less intense diffraction peak is related to the implanted layer (strained state) and is found shifted of about 1° in 2θ toward low diffraction angles.

Micro diffraction measurements

Micro X-ray diffraction experiments have been done at the BM32 beam line of the European Synchrotron Radiation Facility - ESRF (Grenoble, France). Data are recorded using reflection mode and a polychromatic x-ray beam (5-13 keV) with a size of 1×2 µm^2. For each samples, around 700 grains of the implanted layer located at the center of the disk have been probed. Figure 3 shows a typical Laue pattern representative of the UO_2 probed grains. It is characteristic of the UO_2 cubic structure, but each Bragg is in fact split: a main Laue spot and a less intense satellite Laue spot, which are respectively related to the strain free substrate and the implanted layer. Indeed, X-ray penetration depth is about than 5 microns, a value greater than the implanted layer thickness.

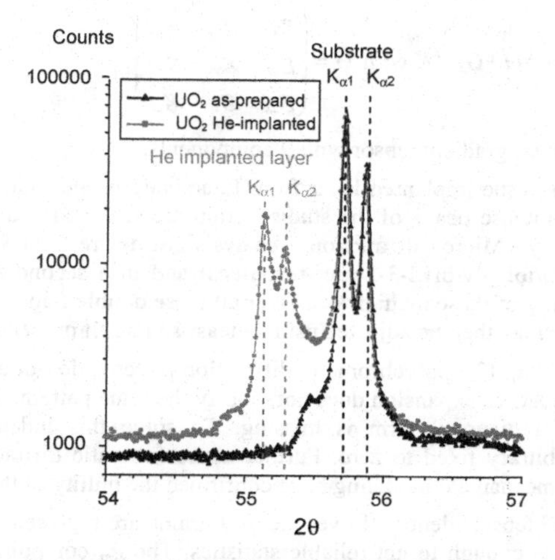

Figure 2 : Diffraction diagrams in θ/2θ mode measured around [311] diffraction peak for as-prepared and implanted samples.

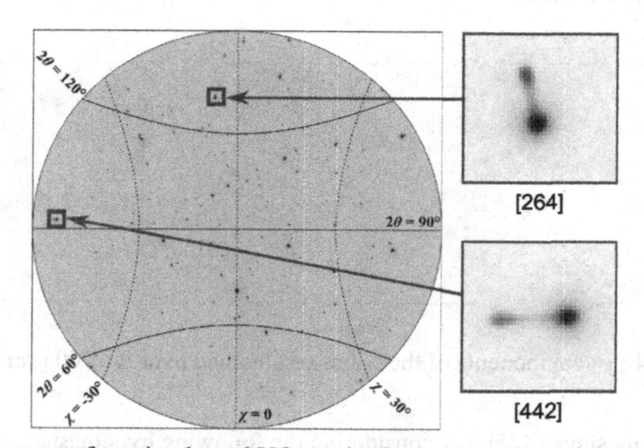

Figure 3 : Laue pattern of one implanted UO_2 grain. Spots are split: an intense diffraction peak due to the strain free substrate and a less intense satellite peak for the implanted layer.

RESULTS AND DISCUSSIONS

Analysis of the micro diffraction data

Strains in the implanted layer lead to a modification of the crystalline lattice. A diffraction vector K in the reciprocal space is then transformed to K' according to the equation:

$$K'=(1+G)^{-1T}K \text{ with } G=\begin{pmatrix} g_{xx} & g_{xy} & g_{xz} \\ g_{yx} & g_{yy} & g_{yz} \\ g_{zx} & g_{zy} & g_{zz} \end{pmatrix} \quad (1)$$

where G is the displacement gradient tensor with 9 components.

For extracting the strain in the implanted layer from Laue pattern, the following procedure has been applied: the more intense peaks of the studied grain are searched and then indexed using XMAS software (X-ray Micro diffraction Analysis Software, ALS Berkeley, USA: https://sites.google.com/a/lbl.gov/bl12-3-2/user-resources) and in a second step, satellite peaks are searched in the vicinity of these main peaks. Over all these double Laue spot sets which have been found, the G tensor may then be adjusted using a least-square fit procedure.

As shown by Ice and Chung [3], polychromatic diffraction patterns do not allow measuring the lattice parameter: a pure lattice expansion does not modify the Laue pattern. The G tensor cannot totally determined i.e. a diagonal term is missing. To solve this indetermination, the g_{xx} component has been arbitrary fixed to zero. Further monochromatic diffraction measurements (macro diffraction) and mechanical modeling have confirmed the nullity of this term.

The g_{ij} components of G tensor identified over the 700 grains are represented in Figure 4. This large number of grains is enough to get reliable statistics. The g_{zz} component amplitude is the more important with an average value of 0.91% and a standard deviation of 0.1%. The g_{xz} and g_{yz} component are on average nil, with nevertheless a strong standard deviation (0.25%). The other components (g_{yx}, g_{zx}, g_{xy}, g_{yy}, g_{zy}) can be considered statistically as nil and their standard deviation (0.05%) as experimental noise.

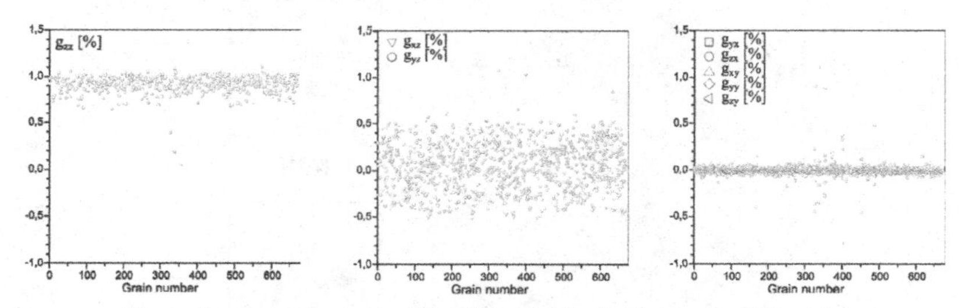

Figure 4 : g_{ij} components of the G tensor obtained over the 700 grains.

A mechanical modeling show [4, 5] that considering the following hypothesis,

- Implanted layer thickness is negligible compared to the sample thickness and small regarding the grain size ;
- Probed grains are located at the center of the sample i.e. far enough from the borders;

the strain in the grains does not depend, in first approximation, on the grain orientation, and its G strain gradient tensor reduced to three terms:

$$G = \begin{pmatrix} 0 & 0 & \varepsilon_5 \\ 0 & 0 & \varepsilon_4 \\ 0 & 0 & \varepsilon_3 \end{pmatrix} \quad (2)$$

where ε_3, ε_4 and ε_5 components describe, following Voigt notation, three components of the strain tensor. It is important to notice that this expression of G corresponds exactly to those experimentally obtained.

Analysis of the macro diffraction data

Macro diffraction in θ/2θ mode allows the measurement of the out of plane strain ε_3. Considering one diffracting plane family [hkl], its value is given by the following formula:

$$\varepsilon_3 = \frac{d_{i,hkl} - d_{0,hkl}}{d_{0,hkl}} \quad (3)$$

where $d_{0,hkl}$ and $d_{i,hkl}$ are the inter reticular distances between [hkl] planes for the strain free substrate and the implanted layer, respectively. The obtained measurements for four different [hkl] planes are reported in table 1. These values are compared to those obtained by micro diffraction. These values are similar and thus reinforce the consistency of the two types (scales) of measurements, validating the choice of a g_{xx} component value equal to zero.

Orientation [hkl]	Macro diffraction ε_3 [%]	Micro diffraction ε_3 [%]
[100]	0.72	0.72
[110]	0.93	0.92
[111]	1.06	1.01
[311]	0.95	0.90

Table 1 : Comparison of the out of plane strain measured by macro and micro diffraction for four [hkl] planes.

Influence of grain crystallographic orientation

According to the model, strains depend at the first order on grain orientation. For studying this influence, ε_3 and $\varepsilon_{45} = \sqrt{\varepsilon_4^2 + \varepsilon_5^2}$ components are represented in Figure 1 using stereographic projection. The different maps obtained by interpolation of the experimental measurements are compared to the results obtained from the model: an excellent agreement between measurements and modeling is found.

These results evidenced the evolution of the strain amplitude in the implanted layer as a function of the grain orientation. It is also important to note that ε_4 and ε_5 components can reach strong amplitude but becomes nil along [100], [110] and [111] principal directions. This is an original result since most of the studies done till now concern single crystals where only these three main directions are investigated [6, 7].

CONCLUSION

Swelling study in a thin implanted layer by He ions at the surface of a UO_2 polycrystal implies several technics: mono chromatic x-ray macro diffraction, poly chromatic x-ray micro diffraction X (Laue mode), and mechanical model. Combining these technics, it is then possible to extract precisely the whole strain tensor. Micro diffraction in the poly chromatic mode is a powerful tool available only on a few synchrotron radiation sources over the world. It allows measurements inside a large number of individual grains with different crystallographic orientations and also intra granular heterogeneities close to grain boundaries for instance. In this study, a mechanical model describing ion implantation at a sample surface has been validated thanks to micro diffraction. The knowledge of the strain gradient can be used to extract the values of the mechanical model from macro diffraction measurements.

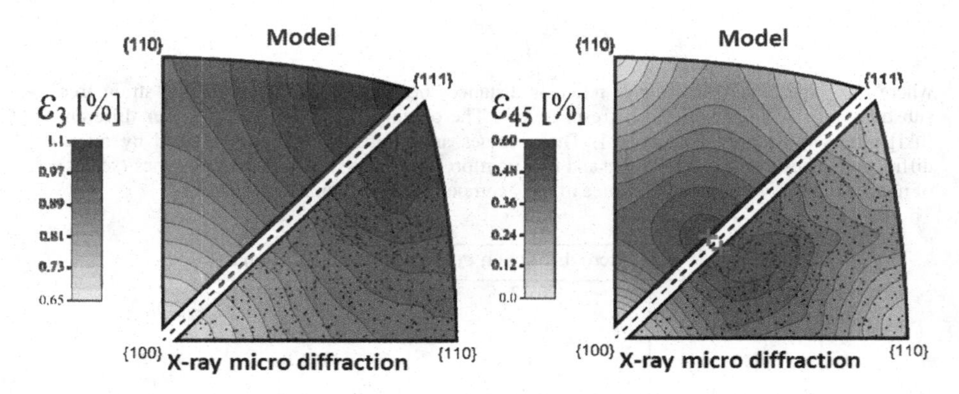

Figure 5 : Comparison between stereographic projections obtained from measurements and calculations (model) for ε_3 and $\varepsilon_{45} = \sqrt{\varepsilon_4^2 + \varepsilon_5^2}$ components.

REFERENCES

[1] S. R. Phillpot, A. El-Azab, A. Chernatynskiy, J. S. Tulenko, JOM 63 (8) (2011)

[2] A. Michel , C. Sabathier , G. Carlot, O. Kaïtasov, S. Bouffard, P. Garcia, C. Valot, Nucl. Instrum. Methods Phys. Res. Sect. B 272, 218 (2012)

[3] J.-S. Chung and G. Ice, Journal of Applied Physics 86, 5249 (1999)

[4] A. Richard, PhD Thesis, Poitiers university, France, 2012

[5] A. Richard, H. Palancher, E. Castelier, J.-S. Micha, M. Gamaleri, G. Carlot, H. Rouquette, P. Goudeau, G. Martin, F. Rieutord, J. P. Piron, P. Garcia, Journal of Applied Crystallography 45, 826 (2012)

[6] A. Debelle, A. Boulle, F. Garrido, L. Thomé, Journal of Materials Science 46, 4683(2011)

[7] W. Weber, Radiation Effects 83, 145 (1984).

Mater. Res. Soc. Symp. Proc. Vol. 1514 © 2013 Materials Research Society
DOI: 10.1557/opl.2013.388

Effect of high temperature heat treatment on the microstructure and mechanical properties of third generation SiC fibers

Dominique Gosset[1], Aurélien Jankowiak[2], Thierry Vandenberghe[3], Maud Maxel[2], Christian Colin[4], Nicolas Lochet[5] and Laurence Luneville[6]

[1]CEA-Saclay, DMN-SRMA-LA2M, LRC CARMEN, 91191 Gif/Yvette, France

[2]CEA-Saclay, DMN-SRMA-LC2M, 91191 Gif/Yvette, France

[3]CEA-Saclay, DMN-SRMA-LA2M, 91191 Gif/Yvette, France

[4]CEA-Cadarache, DER-SRJH-LEDI, 13108 St Paul-lès-Durance, France

[5]CEA-Saclay, DMN-SRMA-LTMEX, 91191 Gif/Yvette, France

[6]CEA-Saclay, DM2S-SERMA-LLPR, LRC CARMEN, 91191 Gif/Yvette, France

ABSTRACT

SiC fibers (High Nicalon S -HNS and Tyranno SA3 -Ty-SA3) submitted to heat treatments in neutral atmosphere up to 1900°C were studied by X-ray diffraction (XRD) and TEM observations then submitted to tensile tests up to 1800°C. The microstructural changes in both materials were determined by XRD using a modified Hall-Williamson method introducing an anisotropy parameter taking into account the high density of planar defects of the SiC-3C structure. HNS fibers exhibit significant modifications in the CDD size which drastically increases from 24 nm to 70 nm in the range 1600°C to 1800°C and in the microstrains which decrease from 0.0015 to 0.0005 between 1750°C to 1850°C. Concerning the Ty-SA3 fibers, no evolution of CDD size and microstrains has been observed. The mechanical properties of single fibers were investigated after the heat treatments showing decreases in the tensile strength reaching up to 20% for Tyranno SA3 and 50% for High Nicalon S. The Weibull moduli were also significantly affected. These results are correlated to the fiber structural and microstructural evolutions.

INTRODUCTION

SiC_f/SiC_m ceramic matrix composites are considered as promising materials for nuclear reactor applications [1,2].Indeed, the third generation of SiC fibers have significantly improved their thermo-mechanical properties due to their near stoichiometric composition, high mechanical strength and thermal stability. Furthermore, these fibers contain a small amount of oxygen (<0.2%) **[3]**. The SiC_f/SiC_m composites exhibit interesting intrinsic features such as a higher operating temperature in comparison to metallic alloys and a low activation level under irradiation. They also provide a very high irradiation stability with a saturation swelling value lower than 0.2% at the aimed operating temperature of about 1000°C [4]. In addition, the fiber/matrix interface which is a key component in these materials significantly increases both the tensile strength and the fracture toughness which can exceed *20 MPa* $m^{1/2}$. It should be mentioned that these properties are not affected by the temperature up to the fiber stability limit.

However, their performances strongly depend on their microstructure which is modified when operating [5,6,7]. In this context, fibers are key components since the mechanical properties of the composite such as creep and tensile strength are greatly influenced by their microstructure [8,9,10,11]. As a consequence, the microstructural stability of the fibers in various environments (i.e. temperature, irradiation, oxidation,...) has to be investigated.

The purpose of this work was to study the mechanical properties for two types of third generation SiC fibers (Hi-Nicalon type S -Nippon Carbon Co. Ltd and Tyranno SA3 -Ube Industry Ltd; respective mean diameter of 14 and 7.5 μm) after heat treatments up to 1900°C and to correlate these results with the observed microstructural evolutions in the same temperature range obtained first with X-ray diffraction (coherent diffraction domains -CDD- and microstrains) and second with transmission electron microscopy [12].

The structure of the fibers consists in a dense packing of equiaxed nanometric β SiC grains resulting in a relative density over 97% [3]. Both fibers contain a small amount of oxygen (<0.2%). Tyranno SA3 fibers exhibit a non uniform concentration in C and Si on the fiber cross section with a larger amount of free carbon in the core of the fiber; they contain aluminum segregating at grain boundaries and a carbon-rich phase is observed at the surface. For Hi-Nicalon type S fibers, free carbon with a turbostratic structure is also found and located between SiC grains.

X-RAY ANALYSIS OF THE FIBERS

It is for long known that X-ray diffraction can usefully be used to obtain microstructural parameters of materials: the analysis of the line profiles of the diffraction patterns directly leads to estimations of the coherent diffraction domains (CDD) size and to the residual microstrains induced by internal defects such as dislocations. In order to allow a correlation analysis of the mechanical properties and the microstructure of the fibers, we resume here results we recently obtained [12].

The X-ray diffraction analyses have been performed on a Bruker D8 Advance diffractometer. The beam is produced with a classical Cu tube (40kV, 40mA) then a Göbel mirror, which leads to a flat, parallel, highly intense, monochromatic ($CuK\alpha_{1+2}$) beam. The detection is made with a multichannel Vantek detector, this allows fast analyses with good angular resolution (channel width = 0.006°). In order to avoid any preferred orientation effect, the fibers are crunched after the heat treatments to short segments (around 50μm long) then put in silica capillaries. The capillary holder allows accurate centering of the capillary and continuous rotation during the analysis.

The diagrams we obtained show monotypic (3C), well crystallized SiC materials. Small grains and high density of defects contribute to high diagrams distortions (Figure 1). Microstructural analyses have been performed with the Hall-Williamson method [12]. We then observed, the distribution of the linewidths correspond to strongly anisotropic materials, this corresponding to a high density of stacking faults or twins along the (111) planes of the SiC-3C structure as shown by TEM observations (Figure 3, left). As compared to the classical Hall-Williamson calculation [13], we then introduced an anisotropy parameter accounting for this planar defect density: the width of a given (hkl) line is made to depend on its direction according to a reference direction. We then use the following expression:

$$\beta.\cos\theta = (\lambda/d + 4.\varepsilon.\sin\theta)\,(1 + \delta\cos^2\chi), \qquad \cos\chi = \frac{h\cdot h_r + k\cdot k_r + l\cdot l_r}{\sqrt{(h^2+k^2+l^2)\cdot(h_r^2+k_r^2+l_r^2)}} \quad (1)$$

with β the integral width (surface over height ratio, corrected for instrumental broadening) of a given (hkl) line, θ its Bragg angle, λ the X-ray (CuKα) wavelength, d the coherent diffraction domain (CDD) size, ε the residual microstrains, δ the anisotropy parameter and χ the angle between the (hkl) and the ($h_rk_rl_r$) reference directions. The best agreement is here obtained with ($h_rk_rl_r$) = (145): Figure 1.This then leads to coherent diffraction domains with apparent platelet shape, the aspect ratio is directly given by δ+1, and anisotropic microstrains.

The main results are reported on Figure 2. The Ty-SA3 fibers show large domains, around *70 nm*, with no modification up to 1800°C and a decrease of the anisotropy parameter above 1600°C. The HNS fibers show small domains, around *25 nm* increasing up to *70 nm* above 1600°C together with a decrease of the anisotropy parameter, but the microstrains relax only above 1750°C, this meaning that two annealing processes are involved. At the highest annealing temperatures, the two materials have the same microstructural parameters values.

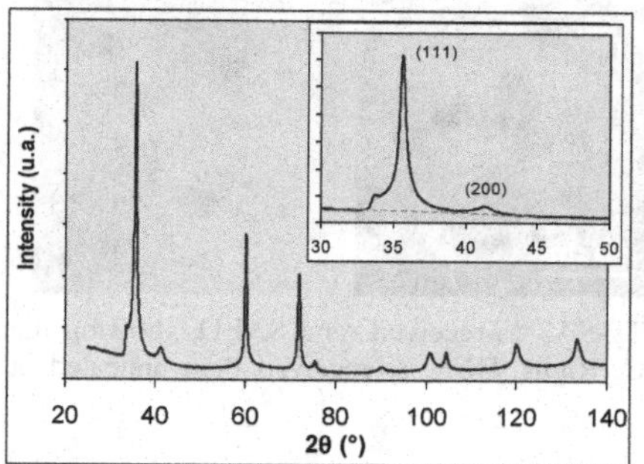

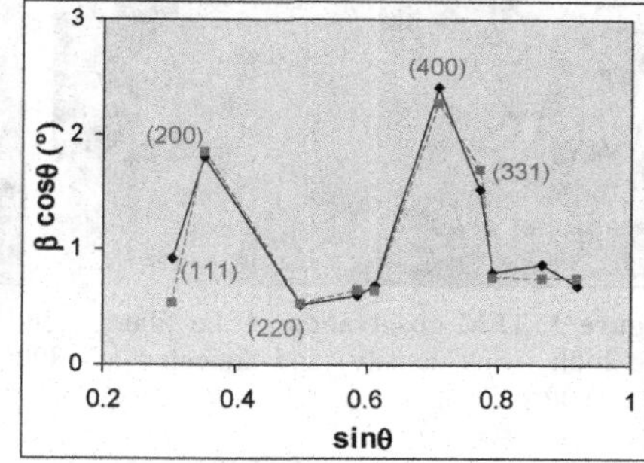

Figure 1. Left: diffraction diagram of the as-received HNS fibers (insert: (111) and (200) lines, evidencing the high faults density).
Right: Hall–Williamson analysis of the diagram.◆: experimental. ■: calculated with eq. 1 and d = *25* nm, ε = *0.0012*, δ = *8.5*.

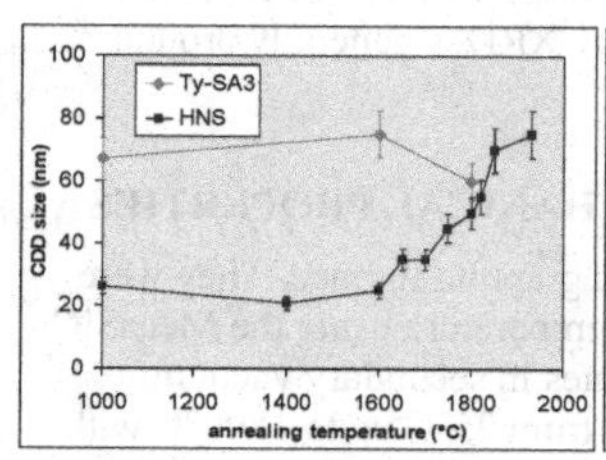

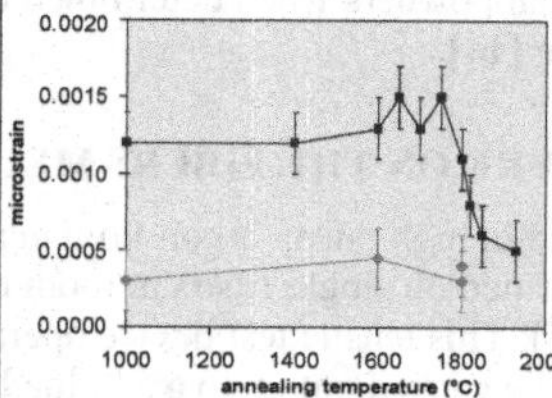

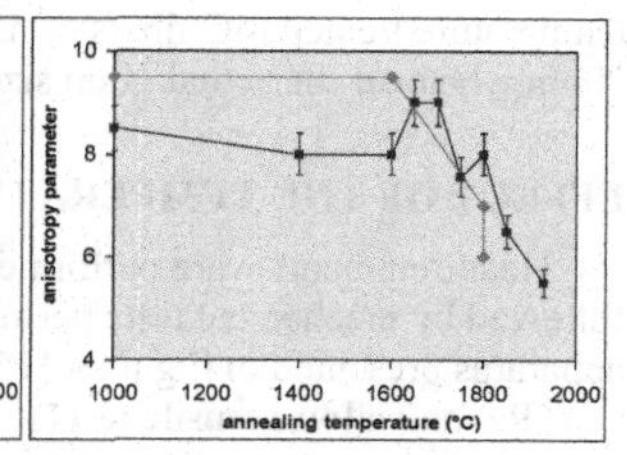

Figure 2. Microstructural parameters (size of CDD, microstrains and anisotropy parameter) of the HNS and Ty-SA3 fibers as a function of annealing temperature.

TRANSMISSION ELECTRON MICROSCOPE OBSERVATIONS OF THE FIBERS

Transmission electron microscope (TEM) observations have been performed in order to check the XRD microstructural parameters estimations [12]. Fibers have been embedded in resin then ion thinned (Precision Ion Polishing System, GATAN 691.056). The observations have been performed on a 200 kV JEOL 2100 TEM with a LaB_6 cathode and a double tilt sample holder stage. The observations coarsely confirm the XRD results: no significant modification of the Ty-SA3 fibers, high grain growth of the HNS fibers. However, the grain size of the Ty-SA3 fibers appears larger than the CDD size. In both cases, a significant decrease of the planar defects density is observed, this meaning the anisotropy parameter is correlated to this faults density.

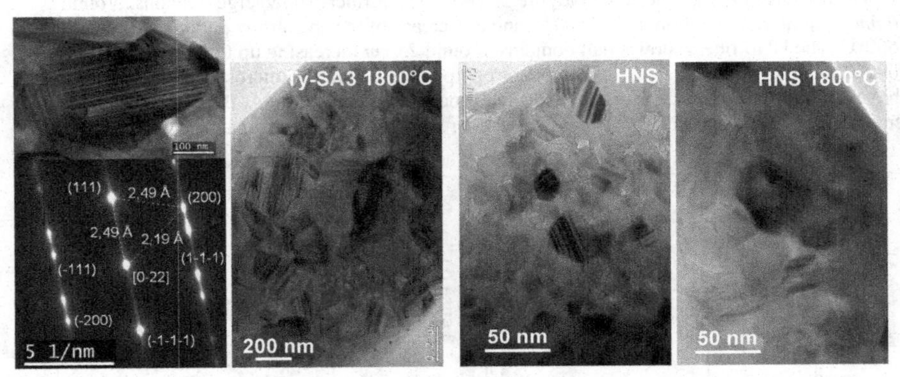

Figure 3. TEM observation of the fibers. Left: Ty-SA3 as-received (and SAED, showing the high twins density) and annealed at 1800°C. Right: HNS, as-received then annealed at 1800°C.

From a comparison of the XRD analysis and the TEM observations, it appears the anisotropy parameter can be related to the planar defects density. On the other hand, the CDD size and the apparent mean grain size are similar for the small grain materials (e.g. as-received HNS). But for the large grain materials, the CDD are smaller than the grain size: this is a quite general result, for the CDD can be bounded by small crystal distortions, CDD size is lower than grain in the sub-micronic range. Other studies lead to the same estimation: the largest CDD size in high temperature treated SiC fibers or nanopowders when determined by XRD is generally around 70nm whatever the actual grain size [14].

EFFECT OF THE TEMPERATURE ON THE FIBERS MECHANICAL PROPERTIES

Heat treatments were performed in high purity argon flow in a graphite furnace. They were followed by mechanical tests performed on single fibers at room temperature using the MecaSiC apparatus presented in Figure 4 [15]. This tensile test device operates in secondary vacuum (10^{-5} Pa) and allows tensile tests to be carried out up to a 5 N load from 25°C up to 1800°C with an electric self-heating of the fiber and using a dedicated LABVIEW program. The apparatus consists in a Melles Griot nanostep 1000-100 motorized linear stage for accurate fiber strain determination combined to a HBM U1A load cell.

Weibull statistics were applied to evaluate the mechanical behavior of the fibers at room temperature. In this study, the as-received fibers were heat treated at 550°C for 30min in helium to remove the fiber sizing. Desized fibers were considered as the reference state for the properties determination. The tensile strength (σ) and Weibull modulus (m) of desized and annealed fibers were investigated. The probability of failure P_r was determined for each tested sample using the classical approach: $P_f = (i-0.5)/n$ with $i = 1 \ldots n$. The m values were deduced from the slopes of the $\ln[\ln(1/(1-P_r)]$ against $\ln(\sigma)$ curves for each set of samples. XRD and TEM analyses have shown that heat treatment induce a fault (planar defects, microstrains) density decrease. In addition, for the HNS fiber, an increase in the CDD size is observed whereas it remains stable for Ty-SA3. From Figure 5, a decrease of 24% in the tensile strength can be deduced for the HNS fibers heat treated at 1600°C. This loss reaches about 50% at 1900°C. Similarly, heat treatments of Ty-SA3 fibers also result in a decrease in the tensile strength. However, the strength loss is only about 13% at 1900°C which is lower than that observed for the HNS fibers.

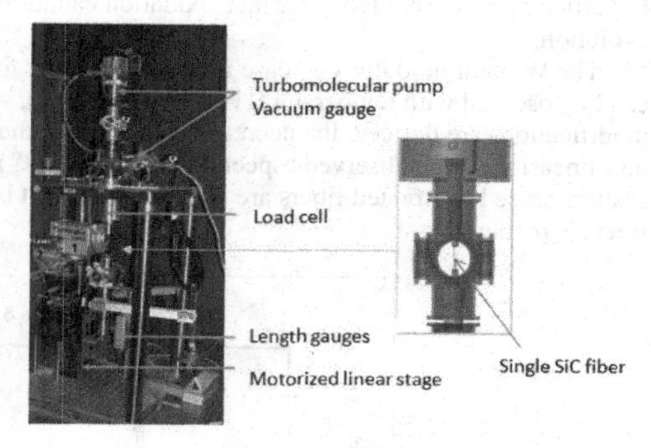

Figure 4. Tensile test device for single SiC fiber

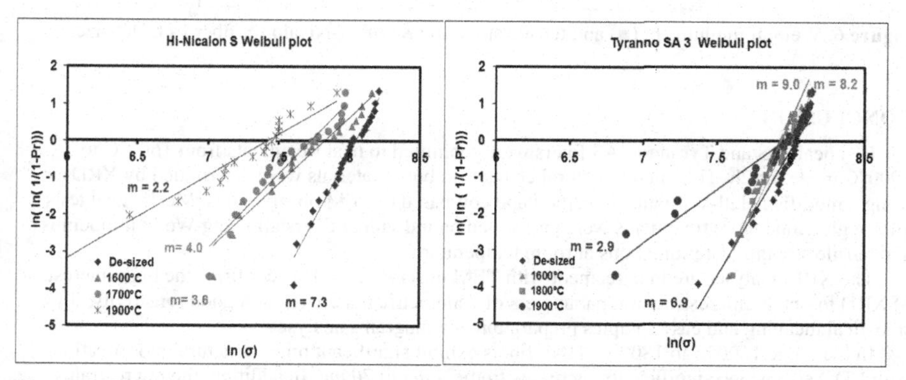

Figure 5. Weibull plots as a function of test temperature. Left: Hi-Nicalon S, Right: Tyranno SA3

As it can be seen in figure 6, for HNS fiber, this tensile strength decrease can be linked to the CDD size increase since this phenomenon initiates at 1600°C and continues at higher temperatures. The Weibull modulus m of Ty-SA3 fibers is also affected by the heat treatments when its microstructural parameters remain nearly constant. This indicates at least another process has to be involved to explain the mechanical properties modifications. It could arise from a modification in the flaws and surface state characteristics occurring during the heat treatment. In particular, flaws related to surface oxidation cannot be excluded to explain the fibers failure evolution.

The Weibull modulus decrease is more important for the HNS fiber where a strong decrease can be observed with temperature. For Ty-SA3 fibers, where only a few microstructural modifications are noticed, the decrease is lower. For the latter, it should be mentioned that some non-linearity can be observed especially in the 1900°C tested fibers. This is mainly due to simple scatters since heat treated fibers are weaker and might be damaged when the specimen (single fiber) is prepared.

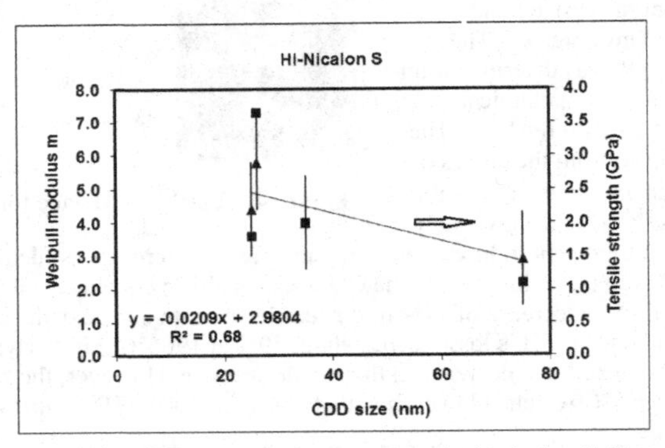

Figure 6. Weibull modulus m (■) and tensile strength (▲) of Hi-Nicalon S fiber vs CDD size.

CONCLUSION

Hi-Nicalon S and Tyranno SA3 fibers were submitted to heat treatments from 1600°C to 1900°C in He for 2h. The microstructural changes in both materials were determined by XRD using a modified Hall-Williamson method and compared to TEM observations. Mechanical tests after high temperature treatments were performed on individual fibers allowing Weibull modulus and tensile strength determinations at room temperature.

The XRD analyses are in agreement with TEM observations. This confirms the high interest of XRD to obtain microstructural parameters of nanometric materials, with good sensitivity, high statistical meaning and easy samples preparation and diagrams analyses.

In the range 1600°C to 1800°C, HNS fibers exhibit significant microstructural modifications with CDD size which significantly increases from *24 nm* to *70 nm*. In addition, the microstrains decrease from *0.0015* to *0.0005* between 1750°C and 1850°C and the planar defects density decreases from 1700°C. Concerning the Ty-SA3 fibers, no evolution of CDD size and microstrains occurred, but the planar defects density decreases in the same temperature range as the HNS fibers. As a consequence, the Ty-SA3 fibers exhibit a more stable microstructure in comparison to the HNS fibers which can explain the Ty-SA3 fibers to have a higher tensile strength after heat treatment. Indeed, the mechanical strength decrease can reach 20% for

Tyranno SA3 and 50% for High Nicalon S for the same heat treatment conditions. The Weibull moduli were also significantly affected. Although a partial correlation between the microstructure evolution and the mechanical properties can be made, other types of flaws are to be involved to explain the two types of fibers to have different mechanical properties after heat treatments at high temperature.

REFERENCES

1 P. Yvon, F. Carré, J. Nucl. Mat., **385-2**, 217–222 (2009)

2 Basic research needs for advances nuclear energy systems, Office of Basic Energy Sciences, U.S. Department of Energy, October 2006

3 C. Sauder, J. Lamon, J. Am. Cer. Soc. **90 [4]** 1146–1156 (2007)

4 W. M. Kriven, Hua-Tay Lin, R. H. Jones, Chapter 38. SiC/SiC Composites for Advanced Nuclear Applications, 27th Annual Cocoa Beach Conference on Advanced Ceramics and Composites: B: Ceramic Engineering and Science Proceedings, Volume 24, Issue 4

5 K. Ozawa *et al.*, J. Nucl. Mat., **367-370** part A, 713-718 (2007)

6 L. L. Snead, O. J. Schwarz, J. Nucl. Mat., **219**, 3-14 (1995)

7 Y Katoh *et al.*, J. Nucl. Mat., **403** [1-3], 48-61 (2010)

8 C. Sauder, J. Lamon, J. Am. Cer. Soc. **90** [4], 1146–1156 (2007)

9 G.E. Youngblood *et al.*, J. Nucl. Mat., **289** [1-2], 1-9 (2001)

10 J. J. Sha *et al.*, J. Nucl. Mat., **329–333** part A, 592-596 (2004)

11 J.J. Sha *et al.*, Mat. Charac., **57** [1], 6-1 (2006)

12 D. Gosset *et al.*, accepted J. Am. Cer. Soc. (2012)

13 G. K. Williamson, W. H. Hall, Acta. Met. **1**, 22-31 (1953)

14 Y. Lee, Y. Park, T. Hinoki, 'Influence of Grain Size on Thermal Conductivity of SiC Ceramics', IOP Conf. Series: Materials Science and Engineering, 18 162014 (2011)

15 C. Sauder, J. Lamon, J. Am. Cer. Soc., **90-4**, 1146–1156 (2007)

Mater. Res. Soc. Symp. Proc. Vol. 1514 © 2013 Materials Research Society
DOI: 10.1557/opl.2013.200

Characterization and thermomechanical properties of $Ln_2Zr_2O_7$ (Ln=La, Pr, Nd, Eu, Gd, Dy) and $Nd_2Ce_2O_7$

Toshiaki Kawano[1], Hiroaki Muta[1], Masayoshi Uno[2], Yuji Ohishi[1], Ken Kurosaki[1] and Shinsuke Yamanaka[1, 2]

[1]Division of Sustainable and Environmental Engineering, Graduate School of Engineering, Osaka University, 2-1 Yamadaoka, Suita, Osaka 565-0871, Japan

[2] Research Institute of Nuclear Engineering, Fukui University, Bunkyo 3-9-1, Fukui-shi, Fukui 910-8507, Japan

ABSTRACT

Pyrochlore type compound $Nd_2(Zr,Ce)_2O_7$ is considered to precipitate in ThO_2-based fuel, that is not observed in irradiated UO_2. In order to evaluate the influences on fuel properties, thermomechanical properties of the pyrochlore type compounds, $Ln_2Zr_2O_7$ (Ln=La, Pr, Nd, Eu, Gd, Dy) and $Nd_2Ce_2O_7$ were investigated. We synthesized the samples by solid-state reaction and pelletized by spark plasma sintering to make high density ($\geq$ 90 %T.D.) pellets. The phase states and lattice parameters were examined by using X-ray diffraction and SEM/EDX analysis. The lattice parameters of $Ln_2Zr_2O_7$ depended on the ionic radii of lanthanide ions. The heat capacity, thermal conductivity, linear thermal expansion coefficient, and elastic constants were also measured. It was confirmed that the thermal conductivities for $Ln_2Zr_2O_7$ were lower than that for ThO_2 and depended on Ln ionic radii. The values of elastic constants tended to increase with increasing the Ln ionic radii, corresponding to the thermal conductivity.

INTRODUCTION

Thorium fuel cycle has many advantages such as higher resistance for nuclear proliferation, less TRU nuclide production and better thermomechanical properties as compared with uranium-plutonium fuel cycle [1,2]. As for fuel form, Thorium dioxide, ThO_2 are considered to be used in existing light/heavy water reactor. In high burnup ThO_2-based fuel, a number of fission products (FPs) are produced in common with UO_2. Several FP elements are known to be soluble in fuel matrix and others form precipitates. These FPs change the physical properties of fuel, therefore it is required to understand the effects of FPs for the utilization of nuclear fuel. However, on ThO_2-based fuel, there are few reports focusing on the FPs effects. To investigate the chemical form of FP elements in ThO_2-based fuel, Ugajin et al. synthesized ThO_2-based SIMFUEL which included simulated FP elements of U, Y, Zr, La, Ce, Nd, Pr, Sr, Ba, Mo, Ru and Pd [3]. In the SIMFUEL, the precipitates, $Nd_2(Zr, Ce)_2O_7$, $(Ba, Sr)MoO_4$ and Ru-Pd alloy were observed. $Nd_2(Zr, Ce)_2O_7$ which has the pyrochlore-type structure have not been found in irradiated UO_2 and $(U, Pu)O_2$. In this study, thermomechanical properties of pyrochlore type compounds $Ln_2Zr_2O_7$ (Ln=La, Pr, Nd, Eu, Gd, Dy) and $Nd_2Ce_2O_7$ which can be formed in the ThO_2-based fuel were evaluated to understand influences of FP precipitates on fuel.

EXPERIMENT

Sample preparation

$Ln_2Zr_2O_7$ (Ln=La, Pr, Nd, Eu, Gd, Dy) and $Nd_2Ce_2O_7$ were synthesized by means of solid-state reactions using powders of ZrO_2, La_2O_3, Pr_2O_3, Nd_2O_3, Eu_2O_3, Gd_2O_3, Dy_2O_3 and CeO_2(purity : $\geq$ 99.9%) as starting materials. The mixture was compacted into pellets at a pressure of 100 MPa for 5 min. These pellets were heated at 1873 K for 48 h in air. Additionally, Ln = Eu, Gd, Dy samples were sintered more 24 hours in a vacuum. The obtained pellets were ground into powder, and then pelletized by spark plasma sintering (SPS) for 10 minutes under Ar flow at 1773~1873 K. Finally, in order to compensate oxygen defect which generate during SPS, they were heated for 24 hours in air at 1273 K.

Measurements

The crystal structure of the samples was analyzed by powder X-ray diffraction (XRD) analysis (Rigaku, Ultima) at room temperature using Cu-Kα radiation. The XRD peaks were corrected by Si external standard and the lattice parameters were calculated by means of Cohen's method. The polished surface of samples was observed by field emission-scanning electron microscope (JOEL, JSM-6500-F) and element distributions were measured by energy dispersive X-ray fluorescence spectrometer (JOEL, EX-64175JMU).

Thermal expansion coefficient of samples was determined by lattice parameters from 300 K to 1073 K by using high temperature XRD. Heat capacity of samples, *Cp* was measured by differential scanning calorimetry (NETZSCH, STA 449C Jupiter) from 313 K to 1073 K. Thermal conductivity was calculated by using the equation $\kappa = \alpha\rho Cp$ where α is the thermal diffusivity, ρ is the sample density. Thermal diffusivity α was measured by a laser flash method (NETZSCH, LFA457) in the temperature range from 300 K to 1473 K in a vacuum. Elastic constant and Debye temperature were measured by ultrasonic pulse-echo method (NIHON MATECH, Echometer 1062).

RESULTS AND DISCUSSION

$Ln_2Zr_2O_7$ (Ln=La, Pr, Nd, Eu, Gd, Dy)

XRD patterns of $Ln_2Zr_2O_7$ are shown in figure 1. From this figure, it is confirmed that all the samples were single phase and any impurity peak was not identified. As for crystal structure, $Ln_2Zr_2O_7$ (Ln=La, Pr, Nd, Eu, Gd) had the pyrochlore structure and $Dy_2Zr_2O_7$ was determined to have the defect fluorite structure judging from exisistance of XRD peaks around 36° and 43° which derive from pyrochlore structure [4]. The pyrochlore structure can be expressed as fluorite structure which consists of two kinds of cations and ordered oxygen vacancies [5]. On the other hand, in the defect fluorite structure, oxygen vacancies are disorderly arranged. It is reported that the ratio of ionic radii for the two cations determines the crystal structure [6].

Figure 2 shows the dependence of lattice parameters on the ionic radius of Ln^{3+} (coordination number = 8) reported by Shannon [7]. The lattice parameter changes linearly with ionic radius of Ln. This result is in good agreement with references [8-11].

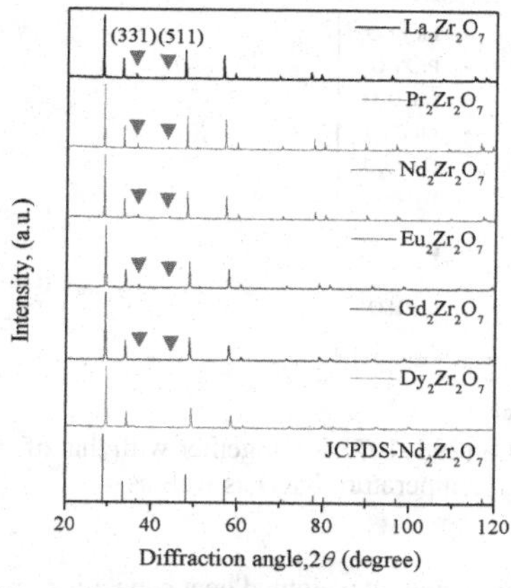

Figure 1. XRD patterns of $Ln_2Zr_2O_7$, together with literature data.

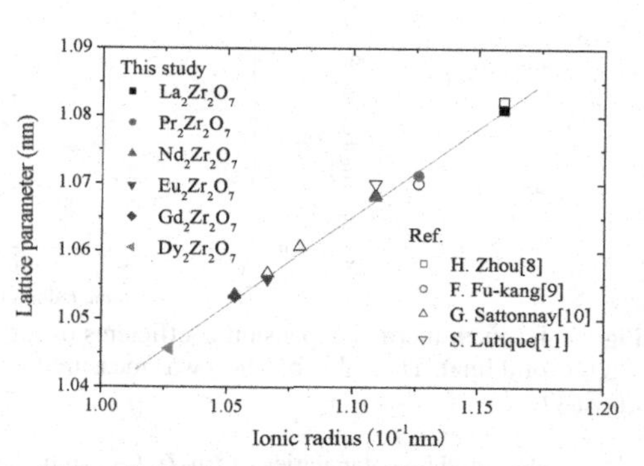

Figure 2. Lattice parameters of $Ln_2Zr_2O_7$, together with literature data [8-11].

SEM/EDS analysis indicated that all the samples were homogeneous and some pores were observed. As a representative, Figure 3 shows SEM image and element distribution of $Nd_2Zr_2O_7$. The relative density of all obtained pellets was over 90 %T.D..

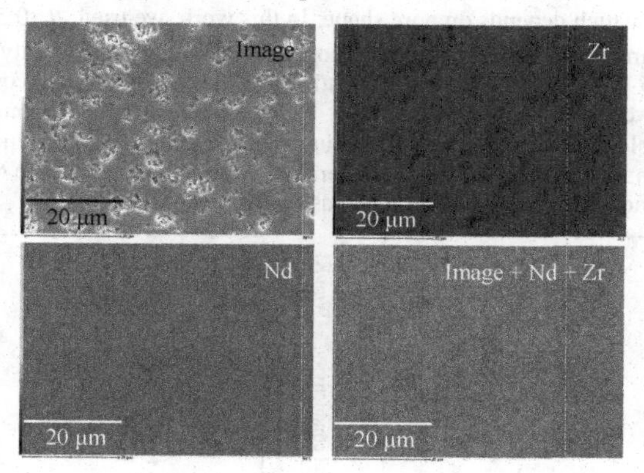

Figure 3. SEM image and element distribution of $Nd_2Zr_2O_7$.

The linear thermal expansion coefficients(298–1273K) of $Ln_2Zr_2O_7$ are shown in figure 4, together with that of ThO_2. All the values of thermal expansion coefficients are nearly equal to that of ThO_2, while those of $Ln_2Zr_2O_7$ roughly increase with decreasing the Ln ionic radius except for $Dy_2Zr_2O_7$. It is assumed that this is because of the difference of crystal structure between pyrochlore and defect fluorite structure.

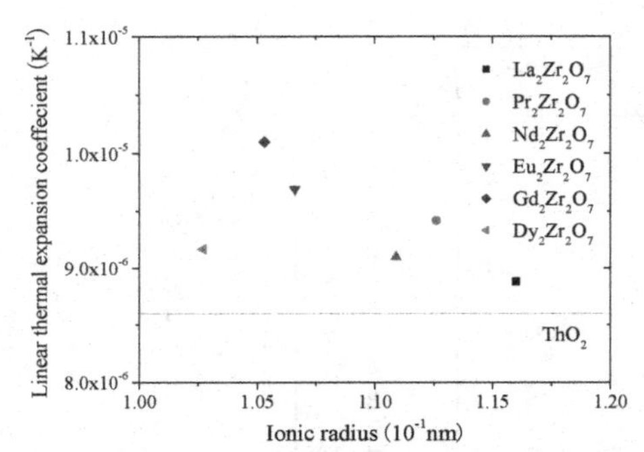

Figure 4. Linear thermal expansion coefficients of $Ln_2Zr_2O_7$ (298–1073 K), together with that of ThO_2 (Solid line). The value of ThO_2 was measured by high temperature XRD as well as $Ln_2Zr_2O_7$.

Measured heat capacities of $Ln_2Zr_2O_7$ can be approximated by weighted heat capacities reported for lanthanide oxide and ZrO_2 [12], using Neumann-Kopp rule. We used those values to calculate the thermal conductivity. The temperature dependence of thermal conductivities for $Ln_2Zr_2O_7$ is shown in figure 5. The thermal conductivity values were corrected to 100 %T.D. by Maxwell-Eucken relation, $\kappa_P = \kappa_0(1-P)/(1+P\beta)$, where κ_P is the thermal conductivity of the porous materials with porosity P, κ_0 is the thermal conductivity of dense materials. β is the correction coefficient which depends on pore shape. In this work, we used β=0.5 value which can be applied to the materials with the spherical pores. The thermal conductivity of ThO_2 is 16.6 W/m/K(@300 K), 5.5 W/m/K(@873 K) , 4.0 W/m/K(@1373 K). The thermal conductivity values for $Ln_2Zr_2O_7$ were much lower than that for ThO_2. Therefore, in high burnup fuel, these precipitates are possible to affect the fuel temperature. The Ln ionic radius dependence of thermal conductivities for $Ln_2Zr_2O_7$ at room temperature is also shown in figure 6. It is indicated that the thermal conductivity decreased with decreasing the Ln ionic radius.

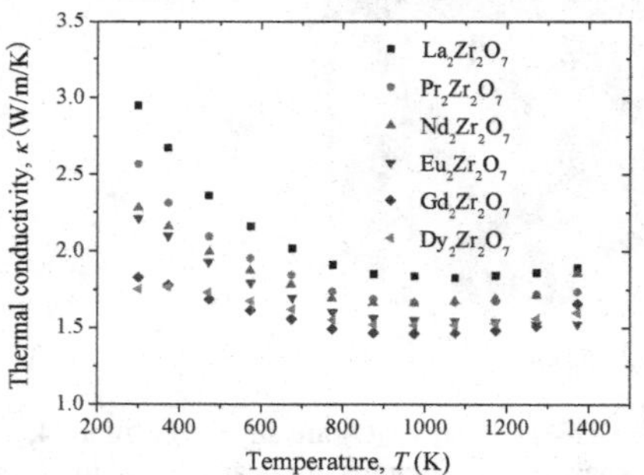

Figure 5. Temperature dependence of thermal conductivities for $Ln_2Zr_2O_7$.

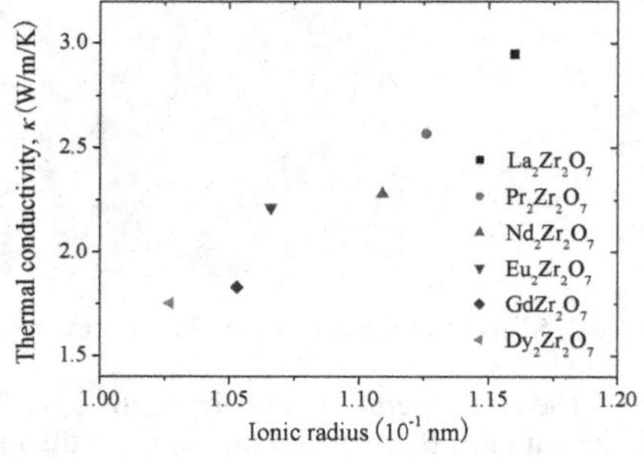

Figure 6. Thermal conductivities for $Ln_2Zr_2O_7$ at room temperature.

Young's moduli and Debye temperatures calculated by longitudinal and transverse sound velocity are shown in figure 7. Each values are corrected to 100 %T.D. based on the empirical equation $E = E_0(1 - AP^n)$, where E is Young's modulus, E_0 is Young's modulus of materials with 100 %T.D., P is the porosity, and A and n are fitting parameters [13]. We determined these fitting parameters by measuring the sound velocity for 5 samples of $Nd_2Zr_2O_7$ with 75~98 %T.D. and applied to $Ln_2Zr_2O_7$. Young moduli and Debye temperatures of $Ln_2Zr_2O_7$ increased with increasing the Ln ionic radii. These results are in agreement with the results of thermal expansion and thermal conductivity qualitatively. It is indicated that the larger differences between ionic radii of cations are, the higher the binding energy become. It can be explained by strain of lattice derived from Ln ionic radius [14]. As compared with ThO_2, the values of Young's moduli of $Ln_2Zr_2O_7$ nearly correspond with that of ThO_2. Debye temperatures of $Ln_2Zr_2O_7$ are higher than that of ThO_2.

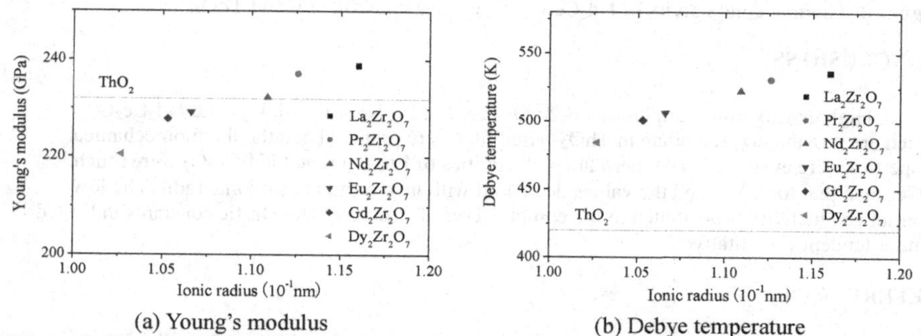

Figure 7. Young's modulus(a) and Debye temperature(b) of $Ln_2Zr_2O_7$, together with that of ThO_2. The solid line indicates data of ThO_2

In spite of the higher Debye temperatures, $Ln_2Zr_2O_7$ indicated lower thermal conductivities than ThO_2. According to the lattice thermal conductivity theory proposed by Slack, the general lattice thermal conductivity can be written as the equation $K_L \propto (M\delta\theta^3)/(n^{2/3}T)$, where M is the average atomic mass of crystal, δ^3 is the average volume occupied by one atom, n is number of atoms per primitive cell and T is the absolute temperature [15]. The value of n for ThO_2 is 3, but that for $Ln_2Zr_2O_7$ is 22 derived from the complex crystal structure. The difference of thermal conductivity is probably due to that for the number of atoms n.

$Nd_2Ce_2O_7$

By XRD pattern and SEM/EDS analysis, obtained $Nd_2Ce_2O_7$ was confirmed to be single phase and had C-type rare earth structure [6]. The lattice parameter was 1.099 nm. The pellet used in measurement was 94.6 %T.D. as is the case of $Ln_2Zr_2O_7$. Thermal conductivity for $Nd_2Ce_2O_7$ is shown in figure 8, together with that for $Nd_2Zr_2O_7$ and ThO_2. Thermal conductivity of $Nd_2Ce_2O_7$ indicates nearly equal value with $Nd_2Zr_2O_7$ and lower value than that for ThO_2. The Young's modulus and Debye temperature were 126 GPa and 360 K respectively. Both values are lower than $Nd_2Zr_2O_7$ and ThO_2.

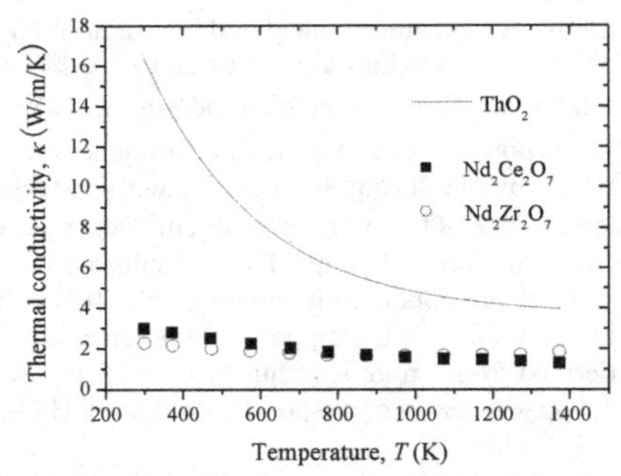

Figure 8. Thermal conductivity of $Nd_2Ce_2O_7$, together with $Nd_2Zr_2O_7$ and ThO_2.

CONCLUSIONS

High density bulk samples of $Ln_2Zr_2O_7$ (Ln=La, Pr, Nd, Eu, Gd, Dy) and $Nd_2Ce_2O_7$ which are possible to precipitate in ThO_2-based fuel were prepared and the thermomechanical properties were evaluated. The thermal conductivities of $Ln_2Zr_2O_7$ and $Nd_2Ce_2O_7$ were much lower than that for ThO_2 and the values decreased with increasing the Ln ionic radii. The low thermal conductivity is attributed to the complex crystal structure. The elastic constants indicated similar tendency qualitatively.

REFERENCES

1. *Potential of thorium based fuel cycles to constrain plutonium and reduce long lived waste toxicity*, IAEA-TECDOC-1349, (2003).
2. *Thorium fuel cycle – Potential benefits and challenges*, IAEA-TECDOC-1450, (2005).
3. M. Ugajin and K. Shiba, *J. Nucl. Mater.* **105**, 211 (1982).
4. K. Shimamura, T. Arima, K. Idematsu, Y. Inagaki, *Int. J. Thermophys*, **28**, 1074 (2007).
5. Francis S. Galasso, "Structure and Properties of Inorganic Solids", Pergamon Press, 1970.
6. H. Nishino, N. Matsunaga, K. Kakinuma, H. Yamamura and K. Nomura, *J. Ceram. Soc. Jpn Supplement 112-1 PacRim5 Special Issue*, **112** [5], S738 (2004).
7. R. D. Shannon, *Acta. Cryst.*, **A32**, 751 (1976).
8. Hongming Zhou, Danqing Yi, Zhiming Yu, Lairong Xiao, *J. Alloys Compd*, **438**, 217 (2007).
9. F. Fu-kang, A. K. Kuznetssov and E. K. Keler, *Bull. Acad. Sci. (USSR) Div. Chem. Soc.*, **4**, 573 (1965).
10. G. Sattonnay, S. Moll, L. Thome, C. Legros, A. Calvo, M. Herbst-Ghysel, C. Decorse and I. Monnet, *Nucl. Instrum. Method*, **272**, 261 (2012).
11. S. Lutique, R.J.M. Konings, V.V. Rondinella, J. Somers, T. Wiss, *J. Alloys Compd*, **352**, 1 (2003).
12. SGTE Pure Substance and Solution Databases, GTT-DATA SERVICE (1996).
13. Rice, "Porosity of ceramics", Marcel Dekker Inc., New York USA (1998).
14. M.A. Subramanian, G. Aravamudan, G.V.S. Rao, *Prog. Solid State Chem.*, **15**, 55 (1983).
15. G. A. Slack in Solid State Physics, edited by F. Seitz, D. Turnbull, and H. Ehrenreich (Academic, New York, 1979), Vol. 34, pp. 1–71.

Mater. Res. Soc. Symp. Proc. Vol. 1514 © 2013 Materials Research Society
DOI: 10.1557/opl.2013.201

High Temperature 2-D Millimeter-Wave Radiometry of Micro Grooved Nuclear Graphite

Paul P. Woskov[1] and S. K. Sundaram[2]
[1] MIT Plasma Science and Fusion Center, 167 Albany Street, Cambridge, MA 02139, U.S.A.
[2] Alfred University, Kazuo Inamori School of Engineering, 2 Pine Street, Alfred, NY, 14802, U.S.A.

ABSTRACT

A dual 137 GHz heterodyne radiometer system was used to study grooved nuclear grade graphite (SGL Group NBG17) inside an electric furnace from room temperature to 1250°C. The millimeter wave radiometer views were collinear with the electric field of one polarized parallel, and the other perpendicular, to the grooves. The anisotropic emissivity was readily detected for 100 μm wide grooves of various depths with a spacing period of 0.76 mm. The emissivity in the 500 – 1250°C temperature range was found to be 5.1 ± 0.5% when the E-field was parallel to the grooves and a factor of 2 – 4 higher, depending on groove depth, in the perpendicular direction. The parallel surface emissivity which was identical to ungrooved surface emissivity corresponded to a 137 GHz surface resistance of 5.3 Ohms, which is about 2.5 times higher than the value predicted from frequency scaling dc surface resistance. The perpendicular emissivity had a modulation with groove depth at odd integral multiples of ¼λ, predicted by electromagnetic finite difference time domain analysis.

INTRODUCTION

Future high temperature nuclear reactors with potential temperature excursions to 1200°C will present new challenges for hot fuel element and structural materials in a high neutron irradiation environment. Graphite, much used for reactors in the past, will continue to be an important material for new reactors because of its very-high-temperature strength, thermal conductivity, and neutron irradiation resistance capabilities [1, 2]. Extending its use into the more extreme environments of future reactors is motivating more studies of graphite properties and the development of new diagnostic tools. A new high temperature measurement capability is needed for resolution of anisotropic properties that can occur either from a material structure that is asymmetric such as a fiber-reinforced matrix composite or from an initial isotropic composition that evolves under non uniform thermal, neutron, and/or structural stresses. For example, fractures will grow approximately linearly in the direction of the least stress [3, 4] causing a dynamically varying anisotropy in the material. Current analysis methods such as x-ray imaging [3] or nondestructive ultrasound [5] are used in laboratory fracture studies, but are not generally applicable to high temperature *in situ* measurements.

Millimeter-wave (MMW) radiometry can provide direction-resolved remote measurement of material emissivity and submillimeter dimensional changes that can be directly correlated with material temperature, resistivity, and swelling [6, 7]. MMWs refer to the electromagnetic wavelength range of 10 – 0.1 mm that can be guided efficiently at high temperature to remote locations. MMW coherent reflection measurements of nuclear graphite at room temperature have been performed [8]. Here we show that MMW passive radiometers can make possible dynamic measurements of anisotropic material properties at high temperature. This approach is demonstrated by MMW emissivity measurnents of grooved nuclear graphite, which suggest that remote monitoring of stress fracturing in graphite is feasible.

EXPERIMENT

A schematic illustration of the laboratory implementation of the MMW thermal analysis hardware is shown in Figure 1. The MMW wave receivers are heterodyne receivers that operate at a center frequency of 137 GHz (λ = 2.188 mm) with intermediate frequency (IF) amplifier bandwidth of 0.5-2.0 GHz making the effective frequency range 137 ±2 GHz. Receiver 2 is polarized perpendicular to Receiver 1 and they are setup in a plane perpendicular to Figure 1 (rotated into the figure plane here for illustrative purposes). Corrugated metallic waveguide (aluminum and brass) direct the MMW signals between the receivers and a steel miter mirror above the furnace where the waveguide material changes to smooth walled ceramic mullite ($3Al_2O_3$:$2SiO_2$) with an inner diameter of 41 mm that goes inside the electric furnace. A Teflon window seals the waveguide section in the furnace to allow a nitrogen purge gas to be flowed through the waveguide around the test specimen and into the furnace to displace the air and prevent uncontrolled graphite sample oxidation at high temperatures. The chopper with a flat reflective copper blade time multiplexes the receiver views into the ceramic waveguide and modulates the signals at about 100 Hz for synchronized detection with lock-in amplifiers. A fused quartz beamsplitter and thermal return reflection (TRR) mirror at Receiver 1 allows an independent check on the sample emissivity [9].

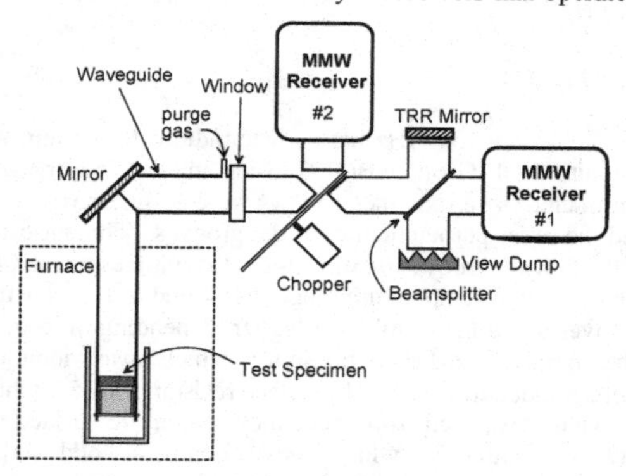

Figure 1. Schematic illustration of the MMW thermal analysis hardware setup

Puck-shaped specimens of 40 mm in diameter by 20 mm thick, as shown in Figure 2a were cut from longer NBG17 graphite rods obtained from the SGL Group. One of the circular flat surfaces was grooved as shown in Figure 2b with 100 μm wide grooves having a spacing period of 0.76 mm. Samples with three groove depths of 1, 1.65, and 2 mm were prepared. They were located for observation in the furnace inside the end of the ceramic waveguide as shown in Figure 1 with the grooved surface facing the radiometers. The MMW temperature observed can be expressed as:

$$T_{mm} = \tau\varepsilon_s T + \tau_c\varepsilon_h T + r\tau\varepsilon_h T \quad (1)$$

where τ is the transmission of the complete transmission line (both hot and cold parts) between the receiver and viewed object, τ_c is the transmission of the cold part of the transmission line, ε_h

Figure 2. a) Photo of graphite puck test specimen (without grooves), 40 mm diameter and b) groove profile used for the measurements (groove depths of 1 and 2 mm also used).

is the emissivity of the hot part of the waveguide, T is the temperature inside the furnace, ε_s is the viewed specimen emissivity and r is the specimen reflectivity, where the emissivity and reflectivity for an opaque object are related by $r = 1 - \varepsilon_s$ through Kirchhoff's law of thermal radiation. The three terms on the right side of Equation 1 can be identified from left to right as the viewed object temperature signal, the direct temperature signal from the waveguide, and waveguide temperature signal reflected by the viewed object. The waveguide parameters can be determined independently and a measurement of the MMW temperature is then used to solve for the MMW emissivity and reflectivity.

DISCUSSION

Figure 3 shows the observed MMW temperatures when the sample with 1.65 mm deep grooves was heated to 525 °C and held at that temperature for about 45 minutes during which calibrations and TRR measurements as indicated were performed. There is a significant difference in the MMW temperature between the two polarizations monitored by Receivers 1 and 2. The graphite was first heated with the grooves oriented parallel to the electric field polarization of Receiver 1, shown by the graph on the left. The Receiver 1 temperature observed was 129°C, while Receiver 2 registered a higher temperature. After cool down the graphite puck was rotated 90° and reheated to the same furnace temperature. This time the Receiver 1 temperature increased to 170 °C while Receiver 2 decreased to the lower temperature as shown on the right. The 30% increase in the observed surface temperature for polarization perpendicular to the grooving over the parallel polarization corresponds to higher actual grooved surface emissivity anisotropy once the contribution from the waveguide emission is in taken into account through Equation 1.

A study was carried out of the grooved depth dependence of emissivity and corresponding surface resistance at temperatures up to 1250°C. The results at 525°C near the graphite resistivity minimum are listed in Table I. The Receiver 1 measurements with the electric field polarization parallel to the grooves were all similar and were the same as for ungrooved surface. This was confirmed by turning over the graphite puck with 2 mm deep grooves and observing its smooth side as listed. The average of all these values gives an

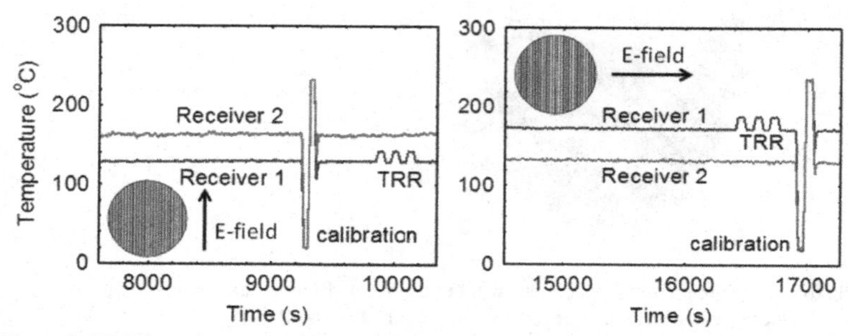

Figure 3. MMW receiver temperatures observed at a furnace temperature of 525 °C viewing a sample of linearly grooved NBG17 graphite. The graph on the right shows the case for Receiver 1 E-field direction parallel to the grooves and the graph on the left shows the case for the E-field of Receiver 1 perpendicular to the grooves as illustrated by the inserts

emissivity of 5.1 ± 0.5%, which can be converted into a surface resistance of 5.3 Ohms through the relation $R_s \approx \varepsilon_s \eta/4$ with higher order terms [10], where η is the impedance of free space (377 Ohms). This surface resistance is higher by a factor of about 2.5 than what would be calculated from the dc measured resistivity [11] of about 7.5 x 10^{-6} Ohm-m at 525 °C (SGL group measurement). This result agrees with a measurement made by Kasparek *et al* [8] at 140 GHz of a graphite grade used in thermonuclear fusion energy experiments. The difference from the dc value can be attributed to microscopic surface roughness that causes the high frequency skin depth to differ from ideal scaling and is commonly observed with other conductors at MMW wavelengths [12].

Table I. NBG-17 Grooved Graphite Anisotropic Emissivity and Surface Resistance at 137 GHz and 525 °C

	Receiver 1			Receiver 2		
Groove Depth	MMW Temperature	Emissivity	Surface Resistance	MMW Temperature	Emissivity	Surface Resistance
[mm]	[°C]		[Ohms]	[°C]		[Ohms]
1	121	0.049	4.7	134	0.112	11.2
1.65	120	0.046	4.5	168	0.195	20.5
2	123	0.054	5.3	139	0.123	12.4
smooth	125	0.058	5.6			
average		0.051	5.3			
error bar		± 0.005	± 0.05		± 0.005	± 0.05

Receiver 2 measurements with the electric field perpendicular to the grooves show a MMW temperature that is hotter, corresponding to grooved graphite surface emissivity and resistance 2 to 4 times higher. The observed normalized increase in surface resistance relative to the parallel groove direction is plotted as circle points in Figure 4 as a function of depth normalized to the MMW wavelength. For good conductors, surface resistance is linearly proportional to emissivity, which in turn is equivalent to absorption loss for an opaque material. The error bars in this plot result from those in the bottom row of Table I. The dashed curve is a theoretical calculation for an aluminum grooved surface at 158 GHz taken from Plaum *et al* [13]. The cycling of the normalized loss through maxima at odd ¼ wavelength multiples of the depth is clearly seen in the experimental and theoretical data. The lower relative loss peak for graphite versus aluminum is likely due to its much lower graphite conductivity, which would decrease a resonance dependent effect. The peaking however is well above the signal to noise ratio of the measurement as evident in Figure 4. This could be exploited in a by a real time monitoring diagnostic to provide depth growth information on developing fractures without an absolute calibration of the instrumentation.

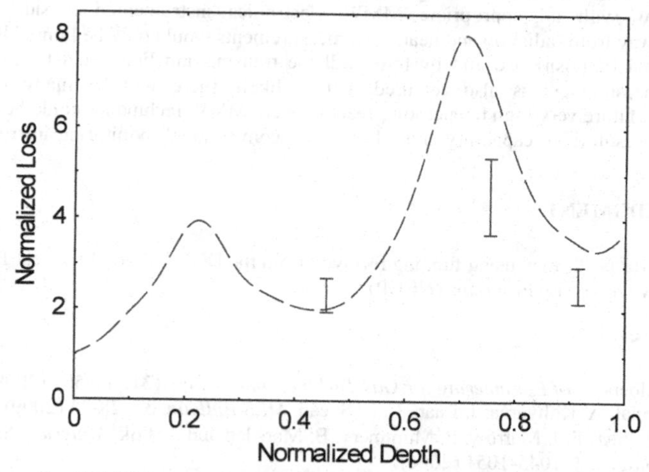

Figure 4. Normalized MMW losses as a function of groove depth normalized to wavelength. Points are for NBG17 graphite data and dashed plot for grooved aluminum surface from a finite difference time domain computation by Plaum *et al* [13].

CONCLUSIONS

MMW radiometry is shown to be an effective tool for observations of hot anisotropic materials, particularly low emissivity materials such as graphite. A heterodyne receiver operating with a polarized waveguide mode is sensitive to viewed material emissivity only in the direction of the electric field. Rotating the direction of the electric field of view would map out material asymmetries. The pair of 137 GHz radiometers used here with collinear views polarized orthogonal to each other has demonstrated the effectiveness of resolving asymmetry imposed on a graphite surface by linear grooving with groove dimensions much smaller than the observation wavelength. SGL Group NBG17 nuclear graphite is found to have an emissivity of ~5% at 137 GHz and 525°C and does not change for electric field of view parallel to the grooves. When the electric field of view is perpendicular to the grooves the emissivity increases by more than a factor of two for groove depths more than 10% of the wavelength and has peaks at odd multiples of ¼λ in depth. The measured ~5% emissivity for smooth surface itself is more than a factor of 2 larger than would be expected by frequency scaling from the dc measured resistivity. This is commonly observed for other conductors in the MMW range suggesting sensitivity to even finer microscopic surface roughness that affects the conductivity skin depth. The measurements also suggest that stress induced fracturing which tends to grow linearly in the direction of least stress could be dynamically monitored in harsh environments to identify stress direction and strength.

MMW radiometry can be extended to measure real-time dimensional changes and fracturing under irradiation as well as thermal annealing. This can be accomplished by accessing and viewing the material surface being irradiated and heated through a long waveguide and MMW window with all appropriate MMW electronics instrumented outside a hostile environment away from radiation and heat. The measurements would only be limited by where a waveguide could be positioned and by how well the transmission line could be calibrated if temperature measurement is also required. It is likely there will be many monitoring requirements in future very high temperature reactors were MMW techniques could be applied to provide new measurement capability and where more conventional monitoring instrumentation may not work.

ACKNOWLEDGMENTS

This research was performed using funding received from the DOE Office of Nuclear Energy's Nuclear Energy University Programs (NEUP).

REFERENCES

1. T. L. Albers, *J. of Engineering for Gas Turbines and Power*, **131,** 064501 (2009).
2. J. P. Bonal, A. Kohyama, J. Laan, L. L. Snead, *MRS Bulletin*, **34**, 28-34 (2009).
3. A. Hodgkins, T. J. Marrow, P. Mummery, B. Marsden and A. Fok, *Materials Science and Technology*, **22**, 1045-1051 (2006).
4. H. Kakui and T. Oku, *J. of Nuclear Materials*, **137**, 124-129 (1986).
5. A.S. Erikssona, J. Mattsson, A.J. Niklasson, *NDT&E International*, **33**, 441-451 (2000).
6. P. P. Woskov and S. K. Sundaram, *Proc. 2010 MRS Fall Meeting*, Symposium R, Boston, (2010).
7. P. P. Woskov, S. K. Sundaram, W. E. Daniel, Jr., D. Miller, *J. of Non Crystalline Solids*, **341/1-3**, 21-25 (2004).
8. W. Kasparek, A. Fernandez, F. Hollmann, and R. Wacker, *Int. J. of Infrared and Millimeter Waves*, **22**, 1695-1707 (2001)
9. P.P. Woskov and S. K. Sundaram, *J. Appl. Phys.*, **92**, 6302-6310 (2002)
10. S. Ramo, J. R. Whinnery, T. V. Duzer, *Fields and Waves in Communications*, 2 ed., p. 290, John Wiley & Sons, New York (1984).
11. Ibid, p. 152
12. P.B. Bharitia and I. J. Bahl, *Millimeter-Wave Engineering and Applications*, p. 202, John Wiley & Sons, New York (1984)
13. B. Plaum, E. Holzhauer, C. Lechte, *J Infrared Milli Terahz Waves*, **32**, 482-495 (2011)

Mater. Res. Soc. Symp. Proc. Vol. 1514 © 2013 Materials Research Society
DOI: 10.1557/opl.2013.139

Reduction of Gd_6UO_{12} for the Synthesis of Gd_6UO_{11}

Darío Pieck [1], Lionel Desgranges [1*], Yves Pontillon [2], Pierre Matheron [3]
[1] *CEA, DEN, DEC, SESC – Laboratoire des Lois de Comportement des Combustibles.*
[2] *CEA, DEN, DEC, SA3C – Laboratoire d'Etudes de la Microstructure des Combustibles Irradiés.*
[3] *CEA, DEN, DEC, SPUA – Laboratoire Combustible Uranium.*
* *Email address: lionel.desgranges@cea.fr, telephone: +33442253159, fax: +33442253285, postal address: CEA, DEN, DEC, Département d'Etudes des Combustibles, SESC/LLCC, bât. 352, F-13108 Saint Paul lez Durance, France.*

ABSTRACT

In the present work, we focus on δ-Gd_6UO_{12} phase and its stability under reducing conditions. This later point is interesting regarding reducing environment that could exist in some nuclear storage sites and that could possibly degrade δ–compounds. A polycrystalline δ-Gd_6UO_{12} sample was prepared by sintering cubic-Gd_2O_3 and UO_2 mixed powders under an air atmosphere. The resulting pellets were then characterized and reduced by heat treatment under an Ar with H_2 5% atmosphere. XRD analysis of the sample after reduction did not confirm the reduction into Gd_6UO_{11} but a decomposition of the δ-compound. Preliminary characterizations of these decomposition products are presented.

INTRODUCTION

One possibility for safe nuclear waste storage consists in their immobilization in ceramics that can resist irradiation damage on large time scales. Compounds with structures similar to fluorite, like δ–compounds, have been proposed for nuclear waste immobilization [1]. Amongst the many studies that have been made concerning radiation damage on such materials, the δ-Gd_6UO_{12} phase which is made of urania and gadolinia mixed in a 3:1 molar ratio, was recently proved not to amorphise under irradiation, even though it undergoes an order-disorder phase transition [2,3]. This compound possesses highly ordered fluorite-related superstructures and is named δ-phase (delta) [2], its structure is rhombohedral and belongs to $R\bar{3}$ space group. Its theoretical density is around 8.1 g cm^{-3} and it presents a yellow pale colour.
Depending on the choice of a storage site, environment conditions in which immobilization matrixes are submitted can vary greatly. In some cases (deep geological storage), the environment can be very reducing compare to matrix fabrication conditions. In this work we are interested on δ-Gd_6UO_{12} stability since a Gd_6UO_{11} phase was reported in the U-Gd-O system but not characterized [4]. In order to evaluate technical feasibility of the δ-Gd_6UO_{12} for nuclear waste immobilization the study of the possible reduction of the compound is mandatory.
In this work, we prepared some δ-Gd_6UO_{12} samples and submitted them to reducing heat treatments.
Many fabrication routes have been reported by various authors [3, 5, 6, 7]. Usually, δ-Gd_6UO_{12} synthesis from mixed powders of urania and gadolinia takes place in two steps: oxidation and δ-

phase formation during a reactive sintering. The first oxidation step can be achieved at 400°C in air atmosphere following the reaction (Eq.1),

$$3UO_2 + O_2 \xrightarrow{\Delta} U_3O_8 \qquad Eq.1$$

In this oxidation process, c-Gd_2O_3 is not affected because it cannot be further oxidized. Nevertheless, higher temperatures could reduce to a half the specific surface of the powders due to a cubic/monoclinic phase transition on Gd_2O_3.

Furthermore, δ-Gd_6UO_{12} appears when the following reaction takes place (Eq.2),

$$U_3O_8 + 9Gd_2O_3 + \frac{1}{2}O_2 \xrightarrow{\Delta} 3Gd_6UO_{12} \qquad Eq.2$$

This transformation occurs at high temperature under an air atmosphere. Rising temperature helps product sintering.

In δ-Gd_6UO_{12}, the U cation has a valence of 6^+. Nevertheless, it's expected that under a reducing atmosphere (e.g. H_2 5%) and high temperatures, the U^{6+} cation will be reduced and Gd_6UO_{11} will be obtained, following (Eq.3),

$$Gd_6UO_{12} \xrightarrow{Ar\,H_2\,5\%} Gd_6UO_{11} + \frac{1}{2}O_2 \qquad Eq.3$$

EXPERIMENTS

A first batch of δ-Gd_6UO_{12} was produced following the technique described by [7]. First, cubic-Gd_2O_3 powders (99.99%, 1.8 m²/g) were mixed by tungsten ball milling in ethanol with UO_2 powders (99.11%, 2.1 m²/g), in a 3:1 molar ratio. After been dried at 60°C, it was granulated and compacted into pellets at 440 MPa. Those pellets were then oxidized in air in a tube furnace at 600°C during 12 h. An olive green colored powder was obtained. XRD measures confirmed the oxidation of UO_2 to α-U_3O_8 and the mere presence of c-Gd_2O_3. Its specific area was 2.8 m²/g. This later product was milled, compacted to 4.94 g cm^{-3} and finally sintered in a tungsten crucible in flowing air at 1500°C for 8 h. XRD characterisation was performed on a polished disk of δ-Gd_6UO_{12} using laboratory apparatus with a wavelength of 1.5405 Å.

A first reduction test under an atmosphere of Ar with 5% H_2 was carried out at 1200°C for 8 h but no changes were observed on the sample. A second test was then performed. Heating rate was 300°C/h. The sample was heated up to 1700°C during 8h and the cooling down rate was 900°C/h. XRD was performed on a polished disk of reduced Gd_6UO_{12} as described for δ-Gd_6UO_{12}.

RESULTS AND DISCUSSION

δ-Gd_6UO_{12} characterisation

Geometric density of δ-Gd_6UO_{12} sintered pellets was 6.35 g cm^{-3}, i.e almost 80% of the theoretical density reported by [8]. Consistently, high porosity was observed by scanning electron microscopy (SEM), Figure 1. In addition, the grain size is relatively small: about 2 µm. SEM observation was difficult because of the non conducting nature of δ-Gd_6UO_{12}. This property disallowed energy-dispersive X-ray spectroscopy.

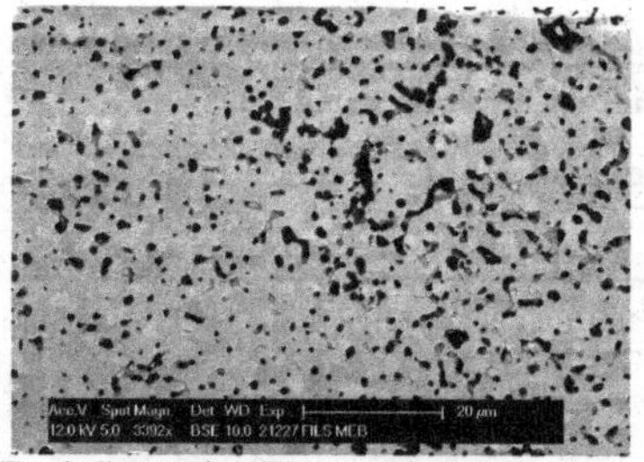

a) Back Scattering Electron image. Relatively small grains are evidenced.

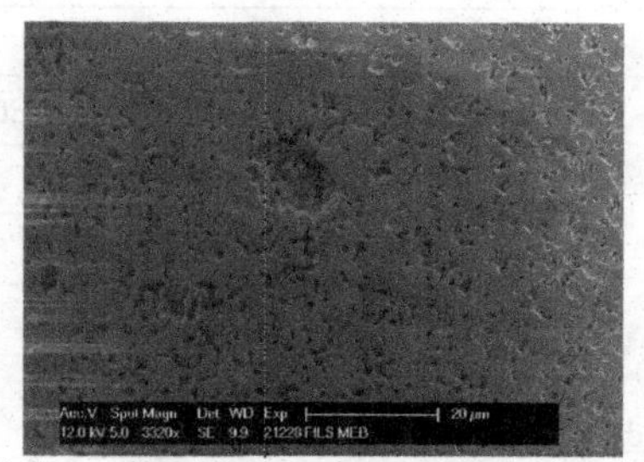

b) Secondary Electron image. High porosity is observed.

Figure 1. Gd_6UO_{12} SEM images.

X-ray diffraction. The XRD pattern is presented in Figure 2. A Rietveld analysis was made based on the reported structure for δ-Y_6UO_{12} [9].

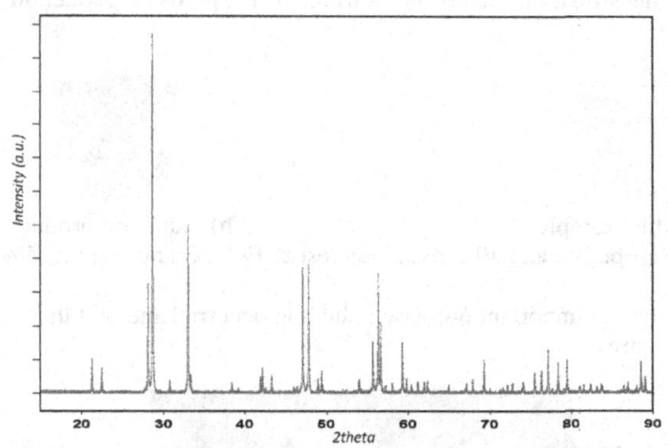

Figure 2. XRD on Gd_6UO_{12} showing only one pure phase. λ_1=1.5405 Å and λ_2=1.5444 Å.

Lattice parameters of the specimen were found to be a = 10.105(6) Å and c = 9.557(8) Å. These values are slightly higher than those reported in literature, e.g. [10] reported a = 10.076 Å and c = 9.529 Å, Table 1.

This difference could be explained considering that delta phase was obtained by a different fabrication route that described in literature. This discrepancy might suggest that the δ-phase could exist in a larger stoichoimetric domain than the pure Gd_6UO_{12}. This hypothesis could explain lattice parameters differences, but can not be confirmed yet. Theoretical density was found to be 8.095 g cm^{-3}.

These results confirmed that our sample is pure δ-Gd_6UO_{12} with characteristic similar to the one reported in literature because it has the same cell parameter and because its crystalline structure is closely related to the previously determined δ-Y_6UO_{12} one.

Table 1. Lattice parameters reported for δ-Gd_6UO_{12}

a	c	Method	reference
a =10.105(6) Å	c =9.557(8) Å	Ethanol milling and air sintering	this work
a = 10.088 Å	c = 9.542 Å	Urea combustion	[8]
a = 10.076 Å	c = 9.529 Å	-	[10]
a = 10.077 Å	c = 9.526 Å	Citrate-nitrate combustion	[11]

δ-Gd_6UO_{12} reduction

No significant evolution of the sample is observed under reducing atmosphere at 1200°C temperature, which implies that some minimum activation energy is needed for reducing processes. During the second reducing test at 1700°C, the initially yellow pale pellet turned black, as it is illustrated in Figure 3. Furthermore the density of the resulting pellet was 7.56 g cm^{-3}, i.e. densification was of 19 % if compared to the starting δ-Gd_6UO_{12} pellet (6.35 g cm^{-3}). This densification could be attributed to porosity reduction and to phase transformation.

a) starting sample **b)** reduction product

Figure 3. δ-Gd_6UO_{12} before and after reduction test at 1700°C under Ar H_2 5% atmosphere.

Indeed, SEM shows that an important porosity reduction occurred and that there was a change of the microstructure, Figure 4.

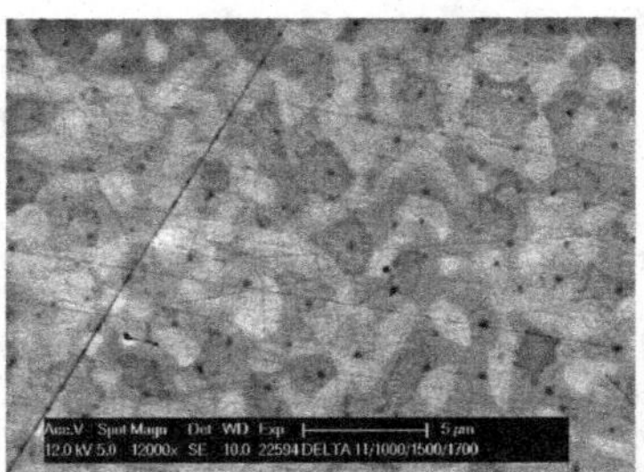

Figure 4. SEM SE image. Reduced Gd_6UO_{12} at 1700°C under Ar H_2 5%. Pellets microstructure is different from the one showed in Figure 1, an important porosity reduction is clearly observed.

High grey contrast in grains colour can be distinguished with Back Scattered Electrons imaging, Figure 5a. Using Energy-Dispersive X-ray Spectroscopy (EDS) it was found that δ-Gd_6UO_{12}

($Gd/(Gd+U) = 0.857$) had decomposed forming grains of $(Gd_z,U_{1-z})_2O_{3+x}$ with an atomic ratio of $Gd/(Gd+U) \geq 0.9$ and a solid solution $(U_{1-z},Gd_z)O_{2-x}$ phase with $Gd/(Gd+U)$ ratio value lower than 0.775.

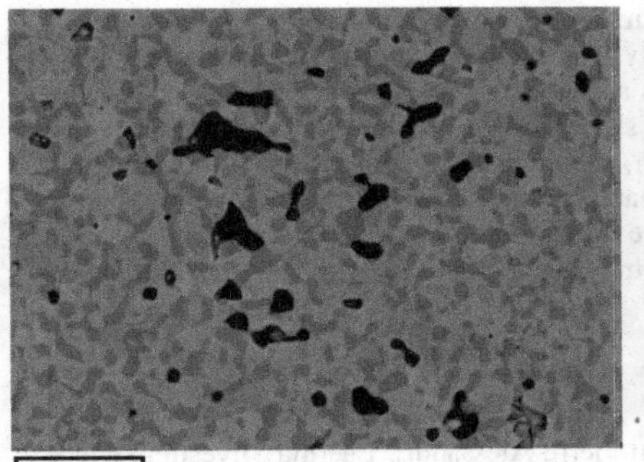

10µm

a) Back Scattering Electron image. High grey contrast in grains colour can be observed.

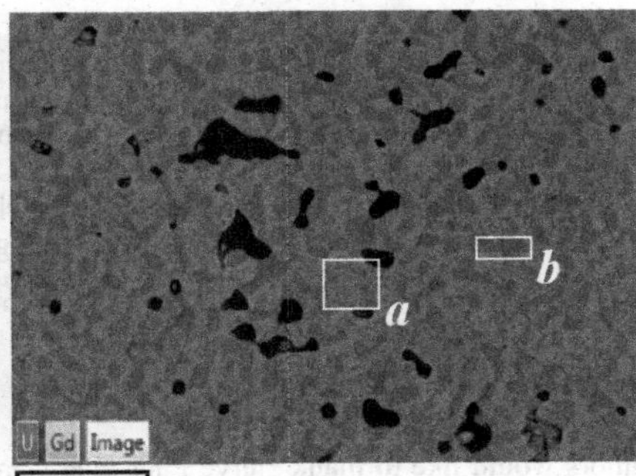

10µm

b) X-ray cartography showing two different chemical compositions. Zone "a" presents a lower Gd content than zone "b".

Figure 5. Reduced Gd_6UO_{12} SEM images.

XRD pattern of the reduced sample at 1700°C is presented in Figure 6. The low resolution of X-ray diffraction did not enable a precise Rietvelt analysis. Nevertheless, diffraction pattern evidenced three different crystalline phases: some remaining hexagonal δ-Gd_6UO_{12} and two newly formed phases. One phase was identified as closely related to m-Gd_2O_3 [12]. The other one was cubic and could correspond to Gd_6UO_{11} although its measured $Gd/(Gd+U)$ ratio is lower than the expected one. Further analyses with better diffraction data are necessary to get a more detailed picture of this phase.

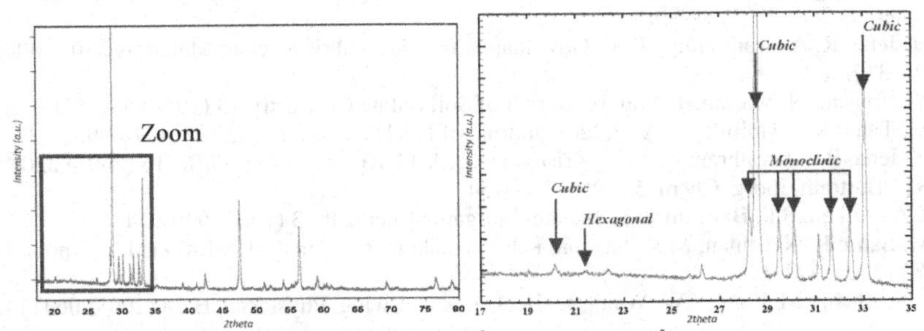

Figure 6. XRD on reduced Gd_6UO_{12}, λ_1=1.5405 Å and λ_2=1.5444 Å.
On the right side figure it can be seem the presence of three phases: some remaining Gd_6UO_{12} (hexagonal) and other two phases: a cubic one and a monoclinic.

CONCLUSIONS

δ-Gd_6UO_{12} has been produced and characterized before and after reduction.
It was found that δ-Gd_6UO_{12} decompose under a reducing atmosphere of H_2 5% at temperatures higher than 1200°C. In the framework of actinide immobilisation in a nuclear storage site, such a decomposition could induce actinide atom mobility inside the immobilisation matrix and could also lead to some actinide release if actinide atoms are insoluble in new phases resulting from δ-Gd_6UO_{12} decomposition. However such a scenario shows high activation energy as evidenced by the high temperature required to observe δ-Gd_6UO_{12} decomposition. More work is still needed to assess the occurrence of such decomposition mechanism at low temperature and over large time scales which are relevant to nuclear waste storage conditions. Next step will be a lixiviation study of this sample in order to evaluate its safety in an accidental scenario of water flooding on a waste storage site.

ACKNOWLEDGMENTS

Authors would like to thank Hervé Palancher, Jean Pierre Alessandri, Laetitia Silvestre, Hélène Rouquette, Claude Verdier, José Sanchez, Catherine Tanguy and Eric Bertrand from CEA Cadarache for their help in experimental work.

REFERENCES

[1] W.J. Weber, R.C. Ewing, C.R.A. Catlow, T.D. de la Rubia, L.W. Hobbs, C. Kinoshita, H. Matzke, A.T. Motta, M. Nastasi, E.K.H. Salje, E.R. Vance, S.J. Zinkle, J. Mater. Res. 13 (1998) 1434-1484.
[2] M. Tang, K.S. Holliday, C. Jiang, J.A. Valdez, B.P. Uberuaga, P.O. Dickerson, R.M. Dickerson, Y. Wang, K.R. Czerwinski, K.E. Sickafus, Journal of Solid State Chemistry 183 (2010) 844-848.
[3] M. Tang, K.S. Holliday, J.A. Valdez, B.P. Uberuaga, P.O. Dickerson, R.M. Dickerson, Y. Wang, K.R. Czerwinski, K.E. Sickafus, Journal of Nuclear Materials 407 (2010) 44-47.
[4] K.W. Song, K. Sik Kim, J. Ho Yang, K. Won Kang, Y. Ho Jung, Journal of Nuclear Materials 288 (2001) 92-99.
[5] H. Jena, R. Asuvathraman, K.V. Govindan Kutty, Journal of Nuclear Materials 280 (2000) 312-317.
[6] Y. Hinatsu, N. Masaki, T. Fujino, Journal of Solid State Chemistry 73 (1988) 567-571.
[7] M. Tang, K.S. Holliday, J.A. Valdez, Journal of Nuclear Materials (2009) 497-499.
[8] H. Jena, R. Asuvathraman, M.V. Krishnaiah, K.V.G. Kutty, Powder Diffr. 16 (2001) 220.
[9] S.F. Bartram, Inorg. Chem. 5 (1966) 749-754.
[10] E.A. Aitken, S.F. Bartram, E.F. Juenke, Inorganic Chemistry 3 (1964) 949-954.
[11] M. Sahu, K. Krishnan, M.K. Saxena, K.L. Ramakumar, Journal of Alloys and Compounds 482 (2009) 141-146.
[12] F.X. Zhang, M. Lang, J.W. Wang, U. Becker, R.C. Ewing, Phys. Rev. B 78 (2008) 064114.

AUTHOR INDEX

SUBJECT INDEX

Printed in the United States
by Baker & Taylor Publisher Services